The
HEARTBEAT
of the WILD

The
HEARTBEAT
of the WILD

DISPATCHES FROM LANDSCAPES
OF WONDER, PERIL, AND HOPE

DAVID QUAMMEN

WASHINGTON, D.C.

Since 1888, the National Geographic Society has funded more than 14,000 research, conservation, education, and storytelling projects around the world. National Geographic Partners distributes a portion of the funds it receives from your purchase to National Geographic Society to support programs including the conservation of animals and their habitats.

Get closer to National Geographic Explorers and photographers, and connect with our global community. Join us today at nationalgeographic .org/joinus

For rights or permissions inquiries, please contact National Geographic Books Subsidiary Rights: bookrights@natgeo.com

ISBN: 978-1-4262-2207-8

Printed in the United States of America
23/MP-PCML/1

To Nick Nichols and Mike Fay

CONTENTS

CONTENTS

FOREWORD

·┼·┼·┼·

The Heart Is a Muscle

Blood doesn't flow through our arteries and veins by gravity or magic or the force of our personalities. It is pushed. What pushes it is an elaborately engineered muscle (or muscular organ) that serves as a pump: the heart. Without the continuing, impelling action of that pump, the rest dies. The heart can survive without a right hand attached to the body in which it beats, or without a left eye, or even without one of the two kidneys; but the kidney or the eyeball or the hand can't survive without the heart. Why am I belaboring this obvious biological fact? For the sake of analogy: The heartbeat of great natural ecosystems can be seen as a combination of features crucial to preserving the viability of the whole, just as auricles and ventricles and valves and innervation are crucial to maintaining the beat of a heart. Foremost among those ecological features are scale, connectivity, diversity, and certain processes. Put them together, and you have a quality that can go by the word "wildness."

Wildness is intangible, maybe even ineffable, but not imaginary. It's a matter of size and function and gloriously unpredictable complexity, as well as a matter of the greenness of the green leaves, the redness of tooth and claw. Religious people might say, *It's like a soul.* Network scientists might say, *It's an emergent property.*

People sometimes imagine that wildness can survive in small or simplified pieces. It cannot. Is a solitary tiger caged in a zoo an example of wildness? I would say no. The poor animal may be frantic and dangerous, but that's different from wild. The zoo may be valuable, as some zoos certainly are; it may teach things about animals to people with no other opportunity for such learning, but it can't easily teach about wildness. Wildness in its fullest sense has got to be big and complex and interactive and uncontrolled by humans. The best zoos can interest people in wildness, can illuminate wildness, can lead people toward wildness, but they can never replace it.

To support these assertions, of course, I've got to define wildness. Not an easy task, and there are multiple possible answers, but here's the best I can do: Wildness is a name we give to living nature, on planet Earth, at its most robust, unfettered, undiminished, dynamic, and diverse.

I've made a few undergirding assumptions. Wildness is biological. The stormy dynamics of Jupiter's Great Red Spot, the eruption of solar flares, and the inexorable suck of a black hole are all formidable phenomena, fearsomely robust and intricate in their ways, but those ways are different from what we unconsciously mean when we say "wildness." They are dynamic but inanimate. Conversely, Times Square at 5:00 p.m. may be wild in its way, and more so on New Year's Eve, and there's plenty of life—the human sort—but not much biological diversity, even if you count the rodents and roaches in the tunnels underfoot. Wildness, as I see it, requires living creatures of many different forms entangled in a system of surging and ebbing interactions, marked by fluctuations that depend on the near-infinite unpredictability of individual behavior as well as the finite predictability of biophysical and biochemical laws. Wildness also assumes a certain threshold of bigness. A piranha in a fishbowl, a single baobab tree that stands in a village roundabout, or a tiger behind bars in a zoo are not wildness. At best, those creatures are small samples from (not of) wildness. They are melon balls scooped from juicy pulp, but they aren't the melon.

What's missing with the caged tiger, the potted piranha, and the solitary baobab are those four crucial ecosystem features: scale, connectivity, diversity, and processes. Each of them is as essential to this property, wildness, as are the animals we tend to think of as its embodiments—animals such as that tiger, that piranha, or the Nile crocodile, the lowland gorilla, the polar bear, the Andean condor, the gray wolf.

What I mean by *processes* are the interactive dynamics of the system, the various activities that turn one form of creature and one form of behavior and one form of matter and one form of energy into the makings of others.

Those processes include photosynthesis, herbivory, pollination, parasitism, competition, predation, seed dispersal, and decomposition, among others. What I mean by *connectivity* are the linkages and dependencies that such processes build among living creatures and their physical environments. What I mean by *scale* is the important reality that connectivity and processes—and yes, biological diversity too—are all dependent on the sheer size of the place where they exist. Big areas of natural landscape can accommodate more kinds of creatures, more connectivity, and more processes than can small areas. This crucial fact was discerned by several observant naturalists over the centuries; delineated in math by a little-known civil engineer, bird-watcher, and amateur ecologist named Frank Preston, in papers published between 1948 and 1962; and then further considered, systematized, and enshrined in theoretical ecology by Robert H. MacArthur and Edward O. Wilson, first in another paper, in 1963, and soon afterward in a landmark book, *The Theory of Island Biogeography,* published in 1967.

You may never have heard of *The Theory of Island Biogeography,* but that book contains some of the key intellectual foundations of the modern conservation movement. Biogeography is the study of what creatures live where. Island biogeography is the study of what creatures live on which islands—and *why.* But the little MacArthur and Wilson volume was about much more than islands, and it awakened a new awareness among ecologists.

The book was about ecosystems as they exist everywhere on the planet: patches of landscape, made into "islands" by the natural arrangement of physical conditions (such as caves and tide pools) and vegetation types (such as gallery forest along a river's bottomland). Insularized ecosystems of that sort obey principles—losing or gaining biological diversity, preserving or surrendering their wildness—very similar to what happens on real islands, water-encircled islands, in the Galápagos or the Hawaiian chain or elsewhere. But there was more: the matter of ecosystem bits made insular by human impacts on landscape. "The same principles apply, and will apply to an accelerating extent in the future," MacArthur and Wilson wrote, "to formerly continuous natural habitats now being broken up by the encroachment of civilization." That's why they wrote the book.

These two scientists foresaw, back in 1967, that human activities (such as logging and plowing and roadbuilding) would increasingly, as our human population and our hungers and our powers grew, leave ecosystems fragmented into insular pieces, surrounded by the oceanic slosh of what we call civilization. That is, surrounded by people and all the accoutrements we require, all the transformations of landscape we commit, replacing biological diversity and connectivity and ecological processes with cities and highways and great universities and operas and libraries and suburbs and driveways and livestock and crops and parking lots and malls and golf courses. MacArthur and Wilson's book analyzed the effects to be expected on such ecological "islands" amid such oceans of human impact, as well as on literal islands amid oceans of salt water.

Foremost of those island effects, the authors recognized, was a decrease in biological diversity related to the size of the island and its distance from the nearest mainland. This had been proven by empirical studies of island faunas, made logical by the work of Frank Preston, and now explained by MacArthur and Wilson in terms of how animals and plants colonize new islands or go extinct from old ones.

The relationship in both cases, scale to diversity and distance to diversity, is inverse: The smaller the island, the fewer species it will contain; the more distant from the nearest mainland, the fewer species will succeed in colonizing it. Diversity is limited because small islands can support only small populations of animals and plants, and small populations face greater jeopardy of extinction. Colonization is limited because distance presents difficulties—animals or plants will be less likely to immigrate successfully to an island very far from the mainland that serves as source. MacArthur and Wilson represented this inverse relatedness with a very simple figure, showing two trend lines crossing each other at the center of a coordinate graph: an immigration line sloping downward with increasing remoteness of the island, an extinction line sloping upward with increasing smallness of the island. Where the lines crossed was a point of balance, an equilibrium representing the expected diversity of that island over time. MacArthur and Wilson called this an "equilibrium model" of island biotas.

Their little book and its equilibrium model, not much noticed when first published, became one of the guiding theoretical works in ecology and conservation—and, eventually, in the discipline that came to be known as conservation biology. Why? Because it did not speak just about islands. It foretold the fate of ecosystems on mainlands, those "formerly continuous natural habitats now being broken up by the encroachment of civilization." It projected the consequences of habitat fragmentation. It warned of the inexorable truth that, as we humans increasingly occupy Earth's surface and arrogate vast areas to our purposes (leaving only smallish fragments of natural landscapes, some of which we call parks or refuges), those fragments will lose biological diversity. And they will lose it in proportion to their smallness and their distance from larger, richer landscapes.

This is why I place scale and connectivity among the four crucial elements of wildness. Diversity, the third, is a blessing dependent largely on those two. Likewise, the fourth, the full presence of ecological processes such as predation, herbivory, competition, and pollination, will exist only where scale and connectivity allow diverse creatures to interact.

Now, MORE THAN half a century later, we live in the future that MacArthur and Wilson foretold. MacArthur did not survive to see it (he died of renal cancer in 1972, much mourned by the ecologists and others who knew him), but Wilson (until he died, also much mourned, on December 26, 2021) did see it, and lamented it eloquently. Those intervening decades of human population growth, consumption, habitat loss, species loss, and ecological fragmentation are what turned Edward O. Wilson into the world's preeminent voice on the conservation of biological diversity.

For me, Ed Wilson has been a lodestar. To speak personally for a moment: I began writing about the natural world 40 years ago, and within half a dozen years I discovered the MacArthur and Wilson book. It guided me toward ideas and drew me to places and issues involving wildness, something I had grown up (in the forests and creek beds of southern Ohio) loving but not much understanding. It leveraged my transition, while I was in my early 30s, from a natural history essayist who had once been a novelist, with almost no formal science education, into a science journalist with a bent for wild places and an author of science books about ecology and evolutionary biology. That little volume, *The Theory of Island Biogeography*, published by Princeton University Press in a drab mustardy cover, probably influenced my life more than any other book I've ever read, with the exceptions of *Absalom, Absalom!* and *On the Origin of Species*.

It led me to faraway forests where I could hear the heartbeat of wildness thumping like blood in my ears: the central Amazon, the Congo Basin, the Okefenokee Swamp, the highlands of New Guinea. It drew me to other places too, where the heartbeat has grown faint and tenuous, its continuation imperiled by incremental losses of just those elements I've mentioned—diversity and connectivity and processes and scale—wondrous but saddening places such as Madagascar. It helped me understand that a great forest, a true forest, cannot exist without, for instance, insects. That a living creek, of the sort I knew in Ohio, cannot be truly alive without salamanders and frogs and newts. That a black-water swamp—not just the gentle and circumscribed Okefenokee, but big, punishing swamps like the ones I walked through with Mike Fay during his Megatransect across the Republic of the Congo and Gabon—will quite properly contain crocodiles or alligators, maybe lungfish or cottonmouths, plus mosquito larvae and mud turtles and leeches. If the swamp doesn't contain those creatures, something is wrong. The vigor of its heartbeat is diminished.

Subtract predators from a great forest, and you diminish it. Subtract amphibians, you diminish it. Reduce its area by 50 percent and cut the rest into fragments with roads, you diminish it. Eliminate the bees and moths and bats, along with their service as pollinators, and you diminish it horribly. Take away scale and connectivity and diversity and the vital processes, bit by bit, and the heartbeat of the wild grows weaker. Eventually it stops. You may still have a stand of trees. But that's a woodlot, not a forest. Your forest is dead. You may have standing water and a few cattails, but your swamp is no longer a swamp. It's a sump.

This book is a selection of my reporting and writing for *National Geographic* magazine over a 20-year period. At the time, these were individual assignments, though forming a pattern that reflected my own interests, what the editors considered appropriate topics for my skills, and the mission of the National Geographic Society. Now, as assembled, they constitute a book that I hadn't originally conceived but that slowly took shape in my mind—a book about people committed to preserving the heartbeat of the wild, about the battles they fight, about the creatures and processes they struggle to protect, and about the places they know and treasure. If I had set out to write that book in 1999, when these efforts began, I could scarcely have found, on my own, the breadth of opportunities and the depth of supporting resources that working for *National Geographic* afforded me. Those editors and I shared one guiding concern, and I'm guessing that you share it too: that the heartbeat of the wild shall not perish from this Earth.

SMALL AUTHORIAL NOTE: Some of these pieces were originally written in the present tense, to give the narratives a feel of immediacy, and almost all were published within several months or a year of being written (timing that counted as fast within the editorial metabolism of *National Geographic*). I have shifted those narratives here into past tense, grounding them within the time when you will read them now. When mentioning distance, length, or weight, I have sometimes spoken in metric units (because scientists generally record data in metric) and sometimes in imperial units (because British and American scientists revert informally to miles, feet, and inches). I've added short prefaces to each piece, and a little updated material after each, to serve as connective tissue and report outcomes of the efforts I describe. In the Megatransect series, there is some slight repetition of expository matter among the three stories—because they tell a single tale but were published as installments in separate issues of *National Geographic,* with a gap of five months between the first and the second, another five months between the second and the third, and I felt a need to refresh readers' memories. I've chosen to retain that bit of overlap here, for the sake of continuity and to suggest the pace and scope of time over which that adventure unfolded.

Spoken comments in quotation marks are verbatim, as captured in my notebook or in a taped interview. Indirect quotes or other summaries of speech, not within quotation marks, are paraphrases, to the best of my memory.

All of these pieces first appeared between October 2000 and August 2020. (Month and year of publication, for each, are given in the back matter.) Twenty years go by fast when you're working hard and having fun.

THE
MEGATRANSECT I

·❦·❦·❦·

The Long Follow

....

In the winter of early 1999, I got a call from a stranger, a genial fellow with a British accent, named Oliver Payne. He was an editor at National Geographic *magazine, he explained, and they had a project in which he hoped I would be interested. An American ecologist, one J. Michael Fay, was going to walk across the last great remaining forests of Central Africa, a thousand miles or more, maybe a year's effort, conducting an exploratory survey of biological diversity—what creatures existed where, and how abundantly—toward the goal of better conserving them. I knew of Mike Fay from his conservation work, knew that he was legendarily tough, serious, and devoid of pretension and formality, but I had never met him. Fay had labeled this expedition, Payne told me, the Megatransect. Hearing the name, I laughed out loud. Vaunting ambition with a scientific flavor, I thought; good for him.*

We would like you to follow Fay, Payne said, and write a series of articles. Not follow him for the whole walk, of course, no, but parts of it. The photography will be done by your friend Nick Nichols.

I grew more interested. Michael (Nick) Nichols is a great photographer, as well as a pal of mine, and we had never worked together. I realized it was he who must have recommended me to National Geographic, *a magazine with which I had never done business. And right then I happened to be very busy, researching and writing a book on big predators. Thanks, but no thanks, I told Mr. Payne.*

Several days later, Mr. Payne called back. We would really like the writer to be you, he insisted. What will it take? he asked. I said I would think about it.

Making the decision, I considered three factors. Did I relish the chance to walk across the Congo with Mike Fay? Yes, indeed. Was the magazine offering me good money? For a magazine writer, yes. Would I have a satisfactory editorial experience—would my writing be treated with care and respect, not mangled, not deprived of all edge and irony, not flattened to safely bland prose and made secondary to the great photography I knew Nick would deliver? Maybe, or maybe not.

But two out of three isn't bad, I thought. Worth the risk. I said yes.

It was the right decision. Oliver Payne turned out to be a keen, deft, and protective editor, as loyal to his writers and the integrity of their work as he was to the institution. Thus began my long relationship with National Geographic *magazine, during which I became devoted to the National Geographic Society as well. And the reporting side of the three Megatransect stories I wrote, walking with Mike, Nick, and our partners (I almost said "on the trail," but there was no trail), came to be one of the great experiences of my life.*

At 11:22 on the morning of September 20, 1999, J. Michael Fay strode away from a small outpost and into the forest, in a remote northern zone of the Republic of the Congo (ROC), setting off on a long and peculiarly ambitious hike. By his side was an aging local man named Ndokanda, a companion to Fay from adventures past and an elder of the Aka people (one of those ethnic groups formerly known as "Pygmy," now an unacceptable term). Ndokanda was armed with a new machete and dubiously blessed with the honor of cutting trail. Nine other Aka crewmen marched after Fay and Ndokanda, carrying dry bags of gear and food. Interspersed among that troop came still other folk—a camp boss and cook, various assistants, Michael ("Nick") Nichols with his cameras, and me.

It was a hectic departure to what would eventually, weeks and months later, seem a quiet, solitary journey. Fay planned to walk across Central Africa, more than a thousand miles, possibly much more, on a carefully chosen route through untamed regions of rainforest and swamp, from northeastern ROC to the Atlantic Ocean on the coast of Gabon. It would take him at least a year. He would receive resupply drops along the way, communicate as needed by satellite phone, and rest when necessary, but his plan was to stay *out there* the whole time, covering the full route in a single uninterrupted push. He would cross a northern stretch of the Congo River Basin, then top over a divide and descend another major drainage, the Ogooué.

Any big enterprise needs a name, and Fay had chosen to call his the Megatransect—*transect* as in cutting a line, *mega* as in mega, a label that variously struck those in the know as amusing or (because survey transects in field biology are generally straight and involve statistically rigorous repetition) inappropriate. Fay was no sobersides, but he hadn't chosen the label to amuse. Behind this mad lark lay a serious purpose—to observe, count, measure, and from those observations and numbers, construct a portrait of great Central African forests before their greatness succumbed to the inexorable nibble of humanity. The measuring began on that September day. One of Fay's entourage, a bright young Congolese named Yves Constant Madzou, paused at the trailhead to tie the loose end of a string to a small tree.

I paused beside him, because I had heard about the string and it intrigued me. In technical lingo, it was a *topofil*. Its other end was enwound on a conical spool inside a Fieldranger 6500, a device of red plastic roughly the size of a canteen, worn on the belt and used by foresters for measuring distance along any walked route. (Personalized GPS units would eventually make such gizmos obsolete, but this was a transitional time, 1999, and Fay as the planner embraced old-school tools as well as the latest technologies, insofar as either could be useful and reliable.) The topofil paid out behind the walking person while the machine counted traversed footage, much as a car's odometer counts traversed miles. Each spool held almost four miles of string. Madzou carried a half dozen extra spools, and somewhere among the expedition supplies were many more. Being biodegradable, the string would quickly disappear down the gullets of termites and other jungle digesters, I'd been told, but the numbers it delivered with such Hansel-and-Gretel simplicity would be accurate to the nearest 12 inches. You couldn't get that precision from a GPS and a map—not at that time, anyway. Running the topofil each day, from this red plastic box on his belt, was one of Madzou's assignments.

Now, as Madzou stepped out after Fay in the first minutes of Day 1, the Fieldranger gurgled in a low, wheezy tone, like an asthmatic old bird dog catching its breath between ducks. Madzou trailed filament like a spider. The string hovered, chest-high, under tension. And I found it pungent to contemplate that, if Fay's expedition proceeded to its fulfillment, a thousand-mile length of string would go spooling through the equatorial jungle. That string seemed an emblem of all the oxymoronic combinations this enterprise embodied—high tech and low tech, vast scales and tiny ones, hardheaded calculation and loony daring, strength and fragility, glorious tropical wilderness and a mitigated smidgen of litter. As he walked, Fay would gather data in many dimensions by many means, including digital video camera, digital audio recorder, digital still camera, notebook and pencil, conductivity meter, thermohygrometer, handheld computer, digital caliper, and hand lens. The topofil would be a quaint but important analog complement to the rest.

Within less than an hour on the first day, we were shin-deep in mud, crossing the mucky perimeter of a creek. "Doesn't take long for the swamps to kick in around here," Fay said cheerily. He was wearing his usual outfit for a jungle hike: river sandals, river shorts, a lightweight synthetic T-shirt that could be rinsed out each evening and worn again the next day, and the day after, and every day after that until it disintegrated. River sandals were preferable to running shoes or tall rubber boots, he had found, because the forest terrain of northeastern ROC was flat and sumpy, its patches of solid ground interlaced with leaf-clotted spring seeps and black-water creeks, each of them guarded by a corona of swamp. A determined traveler on a compass-line march was often obliged to wallow through sucking gumbo, cross a waist-deep channel of whiskey-dark water flowing gently over a bottom of white sand, wallow out through the muck zone on the far side, rinse off, and keep walking. Less determined travelers, in their Wellingtons and bush pants, just didn't get to the places where Fay went.

He stopped to enter a datum into his yellow Rite in the Rain notebook: elephant dung, fresh. Blue-and-black swallowtail butterflies flashed in sun shafts that penetrated the canopy. He noted some fallen fruits of an under-story tree he identified as *Vitex grandifolia*. Trained as a botanist before he shifted focus to do his doctorate on western lowland gorillas, Fay had an impressive command of the botanical diversity upon which big mammals depend; he seemed familiar with every tree, vine, and herb. He knew the feeding habits of the forest elephant (*Loxodonta cyclotis,* the smaller species of African elephant, adapted to the woods and soggy clearings of the Congo Basin) and the life cycles of the plants that produce the fruits it prefers. He could recognize, from stringy fecal evidence, when a chimpanzee had been eating rubber sap. He could identify an ambiguous tree by the smell of its inner bark. He saw the forest in its particulars and its connectedness. Now he bent pensively over a glob of civet shit. Then he made another notation.

"Mmm. This is gonna be fun," he said, and walked on.

MIKE FAY WASN'T the first half-crazed white man to set out trekking across the Congo Basin. In a tradition that includes such Victorian-era explorers as David Livingstone, Verney Lovett Cameron, Pierre Savorgnan de Brazza, and Henry Morton Stanley, he was merely the latest. (There was also Mary Kingsley, white and Victorian but not male, and arguably more doughty, unpretentious, and good-humored than all the other precursors.) Like Stanley and some of those others, Fay had a certain gift for command, a level of personal force and psychological savvy that allowed him to push a squad of men forward through difficult circumstances, using a mix of inspirational goading, promised payment, sarcasm, imperiousness, threat, tactical sulking, and strong example. He was a paradoxical fellow and therefore hard to ignore, a postmodern redneck who chewed Red Man tobacco, disdained political correctness, knew a bit about tractor repair and a lot about software, and viewed the crowded, suburbanized landscape of modern America with cold loathing. Born in New Jersey, raised there and in Pasadena, he saw no going back; he would live out his life and die in Africa, he said. What made him different from those legendary Victorian zealots was that he traveled not in service of God, or empire, or the personal enrichment of the king of Belgium; he traveled in service to the wild places themselves. He did have sponsors, most notably the National Geographic Society, and the Wildlife Conservation Society (WCS) of New York, of which he was a staff member on paid leave, but he certainly wasn't laboring for the greater glory of them. His driving motive—or rather, the first and most public of his *two* driving motives—was to help protect the most biologically diverse and intact remaining forests of Central Africa by documenting what was in them and where.

His immediate goal was to collect a huge body of varied but intermeshed information about the living riches of the ecosystems he would walk through, and about the degree of human presence and impact. He would gather field notes on the abundance and freshness of elephant dung, leopard tracks, chimpanzee nests, and magisterial old-growth trees. He would make recordings of birdsong for later identification by experts. He would register precise longitude-latitude readings every 20 seconds throughout the walking day, with his Garmin GPS unit and the antenna duct-taped into his hat. He would collect rock samples, note soil types, listen for half a dozen different kinds of skrawking monkey. He would detect gorilla presence by smell and by the freshly chewed stems of a creeper vine *(Haumania danckelmaniana)* that gorillas munch like celery. Beyond the immediate goal, his ultimate purpose was to systematize those data into an informational resource unlike any ever before assembled on such a scale—and to see that resource used wisely by the managers and politicians who would make decisions about the fate of African landscapes. "It's not a scientific endeavor, this project," Fay acknowledged during one of our talks before departure. Nor was it a publicity stunt, he argued, answering an accusation that had been raised. What he meant to do, he explained, was to "quantify a stroll through the woods."

Then there was his second driving motive. He didn't voice it explicitly, but I will: Mike Fay is an untamable man who just loves to walk in the wilds.

Completing this marathon trek wouldn't be easy, not even for him. There were dire diseases, minor health hassles, political disruptions (such as the civil war that wracked the Republic of the Congo in 1997), and other mishaps that could stop him. He was familiar with malaria, aware of filariasis and Ebola, and had found himself inconveniently susceptible to foot worms, a form of parasite that can travel from elephant dung into exposed human feet, burrowing tunnels in a person's toes, only to die there and fester. He was aware that every scratch on an ankle or an arm, in this teeming environment, was a potential infection. He had tasted the giddy vulnerability of facing armed poachers unarmed, confiscating their meat, burning their huts, and wondering bemusedly why they didn't just kill him. But the biggest challenge for Fay would come *after* all his walking.

Could he make good on the claim that this encyclopedia of field data would be useful? Could he satisfy the doubters that it wasn't just a stunt? Could he channel his personal odyssey into practical results for the conservation of African forests?

He was very stubborn, I recognized. Maybe he could.

Suddenly, two kilometers on, Fay made a vehement hand signal: *Stop*. As we stood immobile and hushed, a young male elephant appeared, walking straight toward us through the understory. Ndokanda slid prudently to the back of the file, knowing well that a forest elephant, nearsighted and excitable, was far more dangerous than, say, a hungry leopard or a runaway truck. Fay raised the video camera. The elephant, visually oblivious and upwind of our smell, kept coming. The videotape rolled quietly. When the animal was just five yards from him and barely twice that from the rest of us, too close for anyone's comfort, Fay said in a calm voice: "Hello." The elephant spooked, whirled around, disappeared with its ears flapping.

Tusks length, about 40 centimeters, Fay said. Maybe 10 or 12 years old, he estimated. It went into his notebook.

Mike Fay at this time was a compact 43-year-old American, with a sharp chin and a lean, wobbly nose. Behind his wire-rimmed glasses, with their round, smoky lenses, he bore a disquieting resemblance to the young Roman Polanski. Say something that was doltish or disagreeable and he would gaze at you silently the way a heron gazes at a fish. But on the trail, he was good company, a man of humor and generous intellect. He set a punishing pace, starting at daylight, never stopping for lunch or rest, but when there were field data to record in his yellow notebook, fortunately, he paused often. That allowed me to keep pace and make notes of my own.

Fay had first come to Central Africa in 1980, after a stint with the Smithsonian Peace Corps (a scientific variant of the U.S. Peace Corps) doing botany up in Tunisia. He signed for another stint on the understanding that he would go to a new national park in the Central African Republic (CAR), near its borders with Chad and Sudan. The park, known as Manovo-Gounda St. Floris, was then just wishful lines on a map. The lines encircled an area rich with wildlife, in a region over which the CAR government exerted virtually no control. It was a savanna ecosystem, fertile and wild, supporting large populations of elephants, black rhinos, giant eland, kudu, giraffes, roan antelopes, and other big mammals. "A million hectares," Fay told me, "and you're the only white man in those million hectares for eight months out of the year. It was like paradise on Earth." Yet it wasn't so paradisaical when Chadian and Sudanese poachers came to slaughter the elephants. Both his love for Central Africa and his ferocity as a conservationist seemed to be rooted in that place and time.

It was at St. Floris, too, where Fay began to—what's the right phrase? *Go AWOL? Step off the ranch? Disappear into nowhere for long periods?* Let's say *leaven* his more focused scientific work with wildcat exploratory journeys. Because the park's landscape was open and flat, he put his Peace Corps–issue Suzuki 125 trail bike to some unauthorized use. "I decided that the way to really see that place was to take long traverses from one road to another, sometimes 70 or 80 kilometers, across the places where no one had ever been." Too many field biologists, in his judgment, never ventured more than a few kilometers from their base camps. Fay rejected such tethering; he hungered to see the wider scope and the interstitial details. He was restless. He would load the little bike with extra fuel, a patch kit for flats, two weeks' worth of food, and *go*.

WE LEFT CAMP just after dawn, on Day 3, and followed the Mopo River downstream along a network of elephant trails. We were a smaller group now, Nick Nichols and his assistant having backtracked to the start, for related photo work at another site, intending to rendezvous with Fay's march some weeks later. Fay, Madzou, and I set out, with one path cutter and his machete, while the rest of the crew were still eating breakfast, giving the four of us a relatively quiet first look at forest activity. Under a high canopy of *Gilbertiodendron* trees, the walking was easy. (The genus *Gilbertiodendron* contains a couple dozen species of leguminous trees native to tropical Africa, but one of them, *Gilbertiodendron dewevrei,* is often dominant in tropical forest, and that seemed to account for many of the magisterial trees beneath which we walked.) The understory was sparse, as it generally tends to be in these dominant stands of *Gilbertiodendron,* and well trampled by elephant traffic. Later, as we swung away from the river onto higher ground, the *Gilbertiodendron* gave way to a mixed forest, its canopy gaps delivering light to a clamorous undergrowth of brush, saplings, thorny vines, and woody lianas, through which we climbed hunchbacked behind the day's point man. The thickest zones of such early successional vegetation were known in local slang as *kaka zamba,* politely translated as "crappy forest." Today, it was Bakembe, younger and stronger than Ndokanda, who chopped us a tunnel through the *kaka.*

The most devilish of the thorny vines was that creeper, *Haumania danckelmaniana,* mentioned already as a favored gorilla food. Looping high and low throughout the understory, weaving kaka zamba into a tropical briar patch, forever finding chances to carve bloody scratches across unprotected ankles and toes, *Haumania* (as Fay called it) was the bushwhacker's torment. Even a Congo walker as seasoned as he had to spend much of his time looking down, stepping carefully, minimizing the toll on his feet. Of course, Fay would be looking down anyway, because that was where so much of the data existed—scat piles, footprints, territorial scrape marks, masticated stems, grouty tracks left by red river hogs nose-plowing through leaf litter, pangolin burrows, aardvark burrows, fallen leaves, fallen fruit. Fay's GPS told us where we were, while his map and our compasses told us which way to go. There were no human trails in this forest, because there were no resident humans, few visitors, and no destinations.

Fay paused over a pile of gorilla shit, recognizing seeds of the fruit of a certain tree *(Marantes glabra)* as a hint about this animal's recent diet. Farther on, he noted the hole where a salt-hungry elephant had dug for minerals. Farther still, the print of a yellow-backed duiker, one of the larger forest antelopes. Each datum went into the notebook, referenced to the minute of the day, which would be referenced in turn by his GPS to longitude and latitude at three decimal points of precision. Years onward, his intricate database would be capable of placing that very pile of gorilla poop at its exact dot in space-time, should anyone want to know.

When it came time to ford the Mopo, Fay waded knee-deep into the channel with his video camera pressed to his face. Spotting a dark lump against the white sand, he groped for it one-handed, still shooting. "Voilà. A palm nut." He showed me the hard, rugose sphere, smaller than a walnut, light in weight but heavy with import. It was probably quite old, he explained. He had found thousands like this in his years of wading the local rivers, and carbon-dating analysis of a sizable sample revealed them to be durable little subfossils, ranging back between 990 and 2,340 years. Presumably they washed into a stream like the Mopo after centuries of shallow burial in the soil nearby. What made their presence mysterious was that this kind of palm, *Elaeis guineensis,* is known mainly as an agricultural species, grown on plantations near traditional Bantu villages at the fringe of the forest and harvested for its oil. *Elaeis guineensis* palms seemed to need cleared land, or at least gaps and edges, and to be incapable of competing in dense, mature forest. The abundance of ancient oil-palm nuts in the river channels suggested a striking possibility: that a vast population of early Bantu agriculturalists once occupied this now vacant and forested region. So went Fay's line of deduction, anyway. He hypothesized that those proto-Bantus cut the forest, established palm plantations, discarded millions or billions of palm nuts in the process of extracting oil, and then vanished, as mysteriously as the Anasazi vanished from the American Southwest. Some scholars argued that natural climate change over the past three millennia might account for the coming and going of oil palms, the natural ebb and return of forest, but to Fay it didn't make sense. "What makes sense," he said, "is that people moved in here, grew palm nuts, and then died out." Died out? From *what*? He could only guess: maybe warfare, or a killer drought, or population over-shoot leading to ecological collapse, or severe social breakdown resulting from some combination of such factors. Or maybe disease. Maybe an early version of AIDS or Ebola or bubonic plague emptied the region of people, abruptly or gradually, allowing the forest to regrow. There was no direct evidence for this cataclysmic depopulation, but it was a theme that would recur throughout Fay's hike. Meanwhile, he dropped that palm nut into a zip-seal bag.

Just beyond the Mopo, we sneaked up on a group of gorillas feeding placidly in a *bai,* a boggy clearing amid the forest. We approached within 30 yards of an oblivious female as she worked her way through a salad of *Hydrocharis* stems. (Fay was a trained botanist as well as a gorilla expert, as I mentioned, having done his Ph.D. fieldwork on the diets of western lowland gorillas, and he called countless plants by their scientific names, partly because there *were* no common names in English.) Fastidiously, this gorilla snipped off the tender white bases, tossing the rest aside. Her face was long and tranquil, with dark eyes shaded beneath her protrusive brow. The hair on her head was red, Irish red, as it generally is among adult lowland gorillas. Her arms were huge, her hands big and careful. Leaving me behind, Fay skulked closer along the bai's perimeter. When the female raised her head to look straight in his direction, the intensity of her stare seemed to bring the whole forest to silence. For a minute or two she looked puzzled, wary, menacingly stern. Then she resumed eating. Fay got the moment on zoom-lens video. Later he told me that he had frozen every muscle while she glowered at him, not daring to lower the camera, not daring to move, while a tsetse fly sucked blood from his foot.

The video camera, with its sound track for verbal annotations and its date-and-time log, was becoming one of his favorite tools. He shot footage of major trees, posing a crewman among the buttresses for scale. He shot footage of monitor lizards and big unidentified spiders. He shot footage, for the hell of it, of me floundering waist-deep in mud. Occasionally, he did a slow 360-degree pan to show the wraparound texture of a patch of forest. And when I alerted him that a leech had attached itself to one of the sores on his right ankle, he videoed that. Then he handed me the camera, while Madzou burned the leech off with a lighter, so that I could capture the operation from a better angle.

Just before noon, he inspected another fresh mound of elephant dung, poking his finger through the mulchy gobs. Elephants in this forest ate a lot of fallen fruit, but just what was on the menu lately? He picked out seeds of various shape and size, identifying each at a glance, reciting the Latin binomials as he tossed them into a pile: *Panda oleosa, Tridesmos stemon, Antrocaryon klaineanum, Duboskia macrocarpum, Tetrapleura tetraptera, Drypetes gossweileri,* and what was this other little thing? He couldn't remember, *wait, wait . . . oh yeah, Treculia africana.* As I squatted beside him, impressed by his knowledge and scribbling the names, he added: "Of course, this is where you get foot worms, standing in elephant dung like this."

We made camp along a tributary of the Mopo. The crewmen erected a roof beam for the main tarp and a log bench for our ease before the campfire. According to the topofil, Madzou reported, our day's progress had been 33,420 feet. Not a long walk, but a full one. After dark, as Fay, Madzou, and I sat eating campfire popcorn, there came a weird, violent, whooshing noise that rose mystifyingly toward crescendo, and then crested—as, *whoa,* an elephant charged through camp, like an invisible freight train with tusks. Sparks exploded from the fire as though someone had dropped in a Roman candle, and the crewmen dove for safety. Then, as quickly, the elephant was gone. Anybody hurt? No. Dinner was served and the pachyderm in the kitchen was forgotten, just a minor distraction at the end of a typical day on the Megatransect.

FAY SPENT THE late 1980s at a site in southern CAR, gathering data on the resident gorillas. He was particularly curious about their food choices (all gorillas are vegetarian, but their local diets reflect the plant availabilities of a given ecosystem) and their nesting behavior. The western lowland gorilla, like the chimpanzee, is known to build sleeping nests from bent or interwoven branches, and with gorillas those nests are sometimes elaborate. Every gorilla above weaning age makes such a nest, simple or fancy, almost every night. By counting nests, therefore, a biologist can estimate gorilla population density, and from nest counts and other evidence left behind as the animals move, inferences can be drawn about group size, demographic composition, and social organization. In other words, a researcher can learn much without even *seeing* gorillas.

One of the methods Fay used during that dissertation work was a standard line-transect survey, which involved cutting straight trails through his study area, creating a rectilinear grid, and then walking the trails repeatedly to count and plot nests. Fay's study-site grid, spanning floodplain and lowland forest from the Sangha River to a smaller stream that ran parallel, was just over three miles wide. He could march all through it, gathering data as he went, in a day. Another of his methods, which proved more congenial to his disposition, was what he labeled a "group follow."

He hit upon this technique, from necessity, toward the end of his fieldwork period. The gorillas were skittish. They generally fled from any contact with humans—that is, mutual visibility or intrusive proximity. Earlier on, Fay had spent much effort trying to habituate certain gorilla groups to his presence. That was difficult, he found. But if a group of gorillas were followed and *not* contacted, there was no need for habituation. He could stay near the group indefinitely—out of sight, beyond earshot—and leave them none the wiser while he collected data from their abandoned nests, dung, and other residual clues. So he started to shadow them that way.

It required keen tracking skills. Fay enlisted those skills in the person of a brilliant local tracker named Mbutu Clement, a member of the Bambenzele clan, who became his mentor and friend. With Mbutu's guidance, he would follow a group of gorillas discreetly but persistently for all of one day or several, holding back at distance enough (several hundred yards) to keep them unaware of his presence. Among the clues Mbutu used were chewed-upon stems of *Haumania danckelmaniana,* the thorny creeper, which gorillas find toothsome. Because its tissue oxidizes quickly when exposed to air, a freshly gnawed stem of *Haumania* retains its whitish inner color for only about five minutes; after 10 minutes, it has turned black. Fay and Mbutu tried to stay within the five-minute range of a gorilla group without being perceived. Such fastidious tracking allowed Fay to learn what the gorillas had been eating, how many nests they had built, how often they defecated, and what their group size, ages, and gender composition might be, while minimizing the chance that he would spook them.

Near the end of the study, in late 1988, he and Mbutu followed one group for 12 days, dawn to dark each day, resting and eating and walking in synchronic rhythm with the gorillas. From a reading of his eventual dissertation, I gathered that "the 12-day follow," as he called it, was a high point in his academic fieldwork. It was also a foundational bit of experience for what he would later attempt in the Megatransect.

He returned to grad school in St. Louis meaning to write that dissertation, but after a few months he shoved it aside (not to be finished until eight years later) and flew back to Africa, seizing the irresistible distraction of more fieldwork. His new assignment was to do some surveys of forest elephants in northern ROC. He inherited this project from a biologist colleague who had developed the methodology, gotten the grant, and then found himself laid up with a broken back. Fay took over, choosing to focus the survey on three remote, difficult ecosystems: an area near the Gabonese border known as Odzala, a vast swampland to the east known as Likouala-aux-Herbes, and, farther north, a zone of trackless forest between the Nouabalé and Ndoki Rivers.

Teaming up with an adventuresome Congolese biologist named Marcellin Agnagna, Fay set himself the delectable (to him) task of traversing all three areas on foot. Elephant data would be the purpose and the result, but the bush travel would be its own reward. For the Odzala trek, they began at a town called Mbomo. "People were amazed that we were going to just walk from Mbomo to Tshembe, which in a straight line is like 130 Ks across the forest," Fay told me. "The villagers thought we were out of our minds." A year later, he returned for a second survey trek in the Nouabalé-Ndoki area, where he had found such a wonderland of undisturbed forest that it would eventually, after much determined but deft politicking by Fay and others, become one of the Republic of the Congo's most treasured national parks. By 1994, Fay himself was director of this Nouabalé-Ndoki National Park project, on a management contract between the ROC government and the Wildlife Conservation Society. He based himself at a village called Bomassa, on the east bank of the upper Sangha River. Although his administrative duties had grown heavy and his political reach had lengthened, he still slid out for a two- or three-week reconnaissance hike whenever the necessity or the excuse arose. And soon after that he began to brainstorm about applying his leg-power approach on a whole different scale.

His widened perspective came literally from the sky: a hundred feet above the canopy in a Cessna 182. Back in St. Louis he had gotten pilot training, and by 1996 he had found grant money to buy the Cessna. He began flying low-altitude excursions over the ROC, Gabon, and neighboring countries, browsing the landscape as though it were a colorful map on his coffee table, taking himself down to the altitude of parrots and hornbills above areas no road had ever crossed. He logged a thousand hours. He saw the real texture of what was out there—the hidden bais where elephants gathered, the thick groves of Marantaceae vegetation representing bounteous gorilla food, the fishing settlements along small rivers, the poachers' camps secreted in the outback, the Bantu villages, and the great zones of forest where neither settlements, camps, nor villages had yet arrived. "Everything came together because of the airplane," Fay said. "It gave me the big picture." The big picture as he soon sketched it was of a single grandiose hike, complemented with overflights for aerial videography, that would seek to embrace, sample, quantify, interconnect, and comprehend as much of the Central African forest as humanly possible. After more than a year of planning, enlisting collaborators (among whom Nick Nichols was crucial, thanks to his great influence at *National Geographic,* where he was a staff photographer), gathering permissions from governments, selling his vision to sponsors, arranging logistical support, packing, and further flying, he parked the Cessna and started to walk.

ON THE AFTERNOON of Day 5 we entered Nouabalé-Ndoki National Park, crossing the Ndoki River in dugout canoes, then paddling up a deep black-water channel through meadows of swaying *Leersia* grass, and continuing onward by foot. We spent the last hour before sunset walking through a rainstorm so heavy it filled the trail with a cement-colored flood. On Day 7 we skirted the perimeter of Mbeli Bai, a large clearing much frequented by elephants and gorillas. Fay's first glimpse of this bai back in 1990, he told me, was an ugly experience: He found six elephant carcasses, some with their tusks already hacked out, others left to rot until the extraction would be easier. The park hadn't yet been decreed and poaching was rampant. In recent years, the situation was much improved. The park had also brought protection to giant trees of the species most valued for timber, such as *Entandrophragma cylindricum,* informally known as sapelli, one of the premium African mahoganies. Pointing to a big sapelli, he said, "There's something you wouldn't see on the other side of the river"—that is, west of the park, where selective logging had already combed away the most formidable trees. Later he noted a mighty specimen of *Pericopsis elata,* far more valuable even than sapelli. A log of *Pericopsis* that size was like standing gold, Fay said, worth about $30,000 coming out of the sawmill. Spotting another, he changed his metaphor: "If sapelli is the bread and butter around here, *Pericopsis* is the caviar."

We lingered through mid-afternoon with a group of eerily brash chimpanzees, which had gathered at close range to watch *us.* The chimps hooted and gabbled and grunted, perching in trees just overhead, sending down pungent but unmalicious showers of urine, scratching, cooing, thrashing the vines excitedly, ogling us with intense curiosity. One female held an infant with an amber face and huge, backlit orange ears, neither mother nor baby showing any fear. A young chimp researcher named Dave Morgan, who had joined us for this leg of the hike, counted 11 individuals, including one with a distinctively notched left earlobe.

It was a mesmerizing encounter, both for us and for the chimps, but after two hours with them we pushed on, then found ourselves running out of daylight long before we reached a suitable campsite. None of us wanted a night without water. We groped forward in the dark, wearing headlamps now, cutting and twisting through kaka zamba, finally stumbling into a sumpy, uneven area beside a muddy trickle, and Fay declared that this would do. Early next morning we heard chimps again, calling near camp. With Morgan's help, we realized that it was probably the group from yesterday, having tracked us and bedded nearby. Camp-following chimps? Weren't they supposed to be terrified of humans, who commonly hunt and eat chimpanzees throughout Central Africa? The sense of weird and unearthly comity only increased when, on Day 8, we crossed into an area known as the Goualougo Triangle.

At 4:15 that morning I was awake in my tent, preparing for the day's walk by duct-taping over the sores and raw spots on my toes, ankles, and heels. To travel the way Mike Fay travels was hard on the feet, even hard on *his* feet, not because of the distance he walked but because of where and how. After a week of crossing swamps and stream channels behind him, I had long since converted to Fay's notion of the optimal trail outfit—river sandals, shorts, one T-shirt that could be rinsed and dried. But the problem of foot care remained, partly because of the unavoidable cuts, stubs, and slashes inflicted on toes and ankles by the *Haumania danckelmaniana* vine and other hazards, and partly because the sandy mud of Congolese swamps had an effect like sandpaper socks, chafing the skin away wherever a sandal strap bound against the foot. So I had adopted the practice of painting my feet with iodine every night (after a bath in whatever stream was nearby) and every morning, and at the suggestion of another tough Congo trekker, a colleague of Fay's named Steve Blake, using duct tape to cover the old sores and protect against new ones. Between the sores and the duct tape I carefully placed little patches of clean white paper—cut from the smooth tabs you pull off a bandage, but I kept the paper tabs and threw away the bandages—so the tape didn't stick to my raw flesh. This system, which was my own modification of Blake's, held amazingly well through a day of swamp-slogging, and although peeling off the first batch of tape wasn't fun, removal became easier on later evenings when there was no more hair on my feet. Because I had a small roll of supple green tape, as well as a larger roll of the traditional (but stiffer, less comfortable) silver duct tape, I even found myself patterning the colors—green crosses over the tops of the feet, green on the heels, silver on the toes: a fashion statement. If my supplies of iodine and tape could be stretched for another 10 days and my mental balance didn't tip much further, I would be fine.

At 4:30 a.m., I heard Dave Morgan, awake now in the tent beside mine, beginning to duct-tape his feet.

Over breakfast, Fay himself asked to borrow my tape for a few patches on his toes and heels. I gave him the silver, selfishly hoarding the green. Then again we walked.

Demarcated by the Goualougo River on one side, the Ndoki River on another, the Goualougo Triangle was a wedge-shaped area extending southward from the south boundary of Nouabalé-Ndoki National Park. In other words, it was ecologically continuous with the park but not (at that time) part of it statutorily, and isolated from the wider world by the two rivers. Having already made our Ndoki crossing, we entered on solid ground from the park.

The Triangle embraced roughly 300 square kilometers of primary forest, including much excellent chimpanzee habitat, a warren of elephant trails, and an untold number of big sapelli trees, all encompassed within a logging concession held by a company called Congolaise Industrielle des Bois (CIB), the largest surviving timber enterprise in northeastern ROC. With two sawmills, a shipyard, a community hospital, and logging crews in the forest, CIB employed about 1,200 people, mostly in the towns of Kabo and Pokola, along the Sangha River. Although the company had shown willingness to collaborate with WCS on management of a peripheral zone south of the park, especially toward restricting the commercial trade in bush meat (wildlife killed for food) coming out of the forest, tension now seemed to be gathering around the issue of the Goualougo Triangle.

Mike Fay originally hoped to see that wedge of precious landscape included in the park, but when the boundaries were drawn in 1993, the Goualougo was lined out. About the same time, CIB acquired the concession from another logging company that went into receivership. After a half decade of benign inattention, CIB now wanted to move toward logging the Goualougo, or at least to conduct an on-the-ground assessment of the timber resource and the costs of extracting it. That assessment—a *prospection,* in the jargon of francophone forestry—would put a price tag on the Triangle. Meanwhile the company, in a spirit that mixed cooperation with hard-headed bargaining, had invited WCS to do a parallel prospection, theirs to assess the area's biological value. Weeks after returning from the Congo, I heard CIB's position on the Goualougo put by the company's president, Dr. Hinrich Stoll. "You cannot just say, 'Forget about it, it is completely protected,'" he told me by phone from his office in Bremen, Germany. "We all want to know how much it is worth." Once its worth had been gauged, both in economic and biological terms, also in social ones, then perhaps the international community of conservationists and donors would see fit to compensate his company—yes, and the working people of Pokola and Kabo, Dr. Stoll stressed—for what they were being asked to give up.

But that talk of compensation, of balancing value against value, of ransoming some of the world's last ingenuous chimpanzees, came later. As I strolled through the Goualougo with Fay, he turned the day into a walking seminar in forest botany, instructing me or quizzing Madzou and Morgan on the identity of this tree or that. Here was an *Entandrophragma utile,* slightly more valuable but far less common than its congeneric *Entandrophragma cylindricum.* Its fruits resembled blackish yams festooned with wiry little roots, not to be confused with the banana-shaped fruit of another *Entandrophragma* species, *candollei.* And here was still another, *Entandrophragma angolense.* What about that tree there—what is it, Morgan? he demanded. Um, an *Entandrophragma?* Wrong, Fay said, that one's *Gambeya lacourtiana.* Of course, to me these were all just huge hulking boles, 30 feet around, rising to crowns in the canopy so high that I couldn't even see the shapes of their leaves. Morgan and Madzou were earnest students.

THE HEARTBEAT OF THE WILD

Fay was a stern but effective teacher, sardonic one moment, lucid and helpful the next, drawing tirelessly on his own encyclopedic knowledge and love for the living architecture of the forest. Now he directed Morgan's attention to the fine, fissured, unflaky bark of *Gambeya lacourtiana,* which was not to be confused with the more subtly fissured bark of *Combretodendron macrocarpum,* which was not to be confused with . . . a pile of lumber awaiting shipment from Kabo.

The good news from Day 8 was that Fay found no *Pericopsis elata,* no standing gold, no caviar, at least along this line of march in the Goualougo Triangle. The bad news was that there existed in this place an abundance of *Entandrophragma,* CIB's bread and butter. By the time the prospection team arrived to confirm or modify those impressions, Fay himself would be somewhere else, continuing his own singular sort of prospection at his own pace and scale.

From the Goualougo Triangle we made our way upstream along the Goualougo River, crossing back into the park. On the evening of Day 11, we were settled near an idyllic little bathing hole, a knee-deep pool with a sand bottom and a fallen log nearby that served well as a shelf for my bottle of Dr. Bronner's soap. Peeling away my duct-tape socks, after a gentle soak underwater, I felt exquisite relief. I washed my feet carefully, the rest of my body quickly, and then, given the luxury of deep clear water, my hair. I rinsed my shorts and T-shirt, wrung them out, put them back on damp. They would dry as my body dried, at the campfire. It had been a good day, enlivened by another two-hour encounter with a group of fearless chimps. For dinner there would be a pasty concoction known as *foufou,* made from manioc flour and topped with some kind of sauce, plus maybe a handful of dried apricots for dessert. Then a night's blissful sleep on the ground, with clean and iodined feet; then morning and fresh duct tape; then another day's walk. Having fallen into his rhythm, I had begun to see why Mike Fay loved this difficult, unrelenting forest so dearly.

Seated beside the campfire, Fay put Neosporin antiseptic on his ragged toes. Several foot worms had burrowed in there and died, mortally disappointed that he wasn't an elephant. The ointment, as he smeared it around, mixed with stray splatters of mud to make an oily gray glaze. No, he affirmed, there was no escaping foot hassles out here. You just had to keep up the maintenance and try to avoid infection. When necessary, you stopped walking for a few days. You laid up, rested. Let them heal. Waited it out.

So he said. I could scarcely imagine what Fay's feet would have to look like before he resigned himself to that.

At the end of Day 13 we made camp on a thickly forested bench above the headwaters of the Goualougo River, which up here was just a step-across stream. Our distance traversed since morning, as measured by the Fieldranger topofil, was 42,691 feet. Our position was 2° 26.297' N, 16° 36.809' E, which meant little to me but much to the great continuum of data. This day, alas and hoorah, had been my final one of walking with Fay, at least for now. (According to the editorial plan, I would return, months later, to share other legs of the hike.) Tomorrow I would point myself toward civilization, retracing our trail of topofil string and machete cuts to the Sangha River. Morgan and three of the Aka would accompany me.

And Fay? He would continue northeastward to the rendezvous with Nick, then loop down again through Nouabalé-Ndoki National Park before heading out across the CIB logging concessions and the other variously tracked and untracked forests of Central Africa. The Megatransect had only begun: 13 days gone, roughly 400 (Fay thought) to go. Many field notes remained to be taken, many video and audio tapes to be filled, much data to be entered in computers, many miles of topofil to unroll. Then would come the challenge of making it all matter—collation, analysis, politics. When he reached the seacoast of Gabon, Fay had told me, he would probably wish he could just turn around and start walking back.

....

So it began for Mike Fay: the Megatransect. And so it began for me: 20 years of work for National Geographic *as a specialist on wildlife biology, wild places, and conservation. I was lucky that I'd gotten to know Dave Morgan as well as Mike Fay, during this first leg of the expedition, because Morgan would prove to be a stayer, a devoted and field-tough young man destined to mature into a seasoned chimpanzee biologist, and whose chosen place would remain the Goualougo Triangle (working there along with his eventual scientific and personal partner, Crickette Sanz) even while Fay wound his way across the African continent. And I was lucky to meet the Congolese men, Ndokanda and others, who were crucial to Fay's mission. There would be still others filling that role by the time I returned.*

The Goualougo Triangle itself, so vulnerable and tantalizing to commercial forestry, with its abundance of high-value Entandrophragma trees, would be worth further efforts at giving it protection, and worth also therefore another story on my list of assigned tasks for the magazine (see "Goualougo and Friends," pages 97–107). In the meantime, Mike Fay kept walking. And I went home to heal the wormholes and other sores on my feet.

THE MEGATRANSECT II

·◆·◆·◆·

The Green Abyss

....

Back in the cold, dry air of Montana, my feet did heal. For months, I heard only indirectly how things were going for Fay. Kathy Moran, the veteran photo editor at National Geographic *responsible for that side of the stories, stayed intermittently in touch with both Mike (on the march) and with Nick (who was now working from a fixed camp along the route, shooting wildlife photos of a sort that required long days of stakeout in a tree platform overlooking a bai) by satellite phone. Kathy sent them occasional care packages of spirit-bolstering treats—chocolate chip cookies, popcorn—to be delivered along with the more pragmatic resupply drops, managed by Tomo Nishihara, Fay's logistics man, based in Libreville, Gabon. The news from Tomo and Kathy reached Ollie Payne and me. During the early months, what I heard was: very tough going. Fay and his team had plunged into a vast swampy area he called the Green Abyss.*

He plodded through that, and I rejoined him at a river crossing toward the southwest edge of the abyss, giving me the opportunity to get just a taste of it.

For this second story, I made two separate trips to Africa, rendezvousing with Fay at resupply points and sharing two very different stretches in the middle of the march. That happened because Ollie had told me, before we started, that three stories were wanted and that, as an assigned writer traveling between continents, I would be allowed and budgeted (as standard at that stage in National Geographic's *history) to fly business class for all three trips. I asked: Could I fly coach and make four trips? I was indulged.*

The second trip for this story gave me the opportunity to walk with Fay through the Minkébé forest block. I didn't want to miss that, because I knew it was a place where Ebola virus lives, and I had become fascinated by the ecological mystery of that dangerous creature.

....

It takes a hardheaded person to walk 2,000 miles across west-central Africa, transecting all the wildest forests that remain between the northeastern corner of the Republic of the Congo and the Atlantic. It takes a harder head still to conceive of covering that terrain in a single, sustained, expeditionary trudge. There are rivers to be ferried across or bridged with logs, swamps to be waded, ravines to be traversed, vast thickets to be carved through by machete, and one tense national border, as well as some lesser impediments—thorny vines, biting flies, stinging ants, ticks, vipers, tent-eating termites, and the occasional armed poacher. As though that isn't enough, there's a beautifully spooky forest about midway on the route that harbors Ebola virus, cause of lethal epidemics in nearby villages during the late 1990s. The logistical costs of an enterprise on such a scale, counting high-tech data-gathering gizmos and aerial support, runs to hundreds of thousands of dollars. The human costs include fatigue, hunger, loneliness, tedium, some diseases less mysterious than Ebola, and the inescapable nuisance of infected feet. It takes an obdurate self-confidence to begin such a journey, let alone finish. It takes an unquenchable curiosity and a monomaniacal sense of purpose.

J. Michael Fay was as obdurate and purposeful as they come. But even for him there arrived a moment, after eight months of walking, when it looked as if the whole venture would end sadly. One of his forest crew, a young Bambenzele man named Mouko, lay fevering on the verge of death. Hepatitis was taking him down fast.

Mouko's illness was only the latest travail. Within recent days Fay had been forced to backtrack around an impassable swamp. He had sent his original Aka crew back to Bomassa, after the early days of the expedition, and gathered a fresh team, but now even they, the healthy ones as well as Mouko, were exhausted and ready to quit. The border crossing from ROC into Gabon, which loomed just ahead, had begun to appear politically problematic—no Gabonese visas to be had for a gang of Congolese Bambenzele. And then a certain trader, an Arab outsider dealing in gold and ivory, went missing between villages along one of the few human footpaths with which Fay's route converged. As authorities reacted to the man's disappearance, Fay began dreading the prospect that he and his feral band might come under suspicion and be sidetracked for interrogation. Suspending the march to nurse Mouko, he found himself stuck in a village with bad water. He was running short of food, with not even enough pocket money to buy local bananas. The Megatransect was in megatrouble.

If Mouko dies, Fay thought, it was probably time to roll up the tents and capitulate. He would abandon his dream of amassing a great multidimensional filament of forest survey data, continuous both in space and in time. He would stop recording all those little particulars—the relative freshness of every pile of elephant dung, the location of every chimp nest and aardvark burrow, the species and girth of every big tree—in the latest of his many yellow notebooks. He would stop walking. Human exigencies would preempt methodological imperatives and vaulting aspirations. If Mouko dies, he figured, I'll drop everything and take the body home.

EVEN FROM THE start, in late September 1999, it had looked like a daunting endeavor—far too arduous and demented to tempt an ordinary tropical biologist, let alone a normal human being. But Fay wasn't ordinary. By his standards, the first three months of walking were a lark. Then the going got sticky.

Having crossed Nouabalé-Ndoki and that stunning wedge of pristine forest known as the Goualougo Triangle, having hiked south through the trail-gridded timber concessions and boomtown logging camps of the lower Ndoki watershed, Fay and his team angled west, toward a zone of wilderness between the Sangha and Lengoué Rivers, both of which drain south to the main stem of the Congo River. What was out there? No villages, no roads. On the national map it was just a smear of green. Fay traveled along elephant trails when possible, and when there were none, he bushwhacked, directing his point man to cut a compass-line path by machete.

A strong-armed and equable Bambenzele named Mambeleme had laid permanent claim to the point-man job. Behind him walked Fay with his current yellow notebook and his video camera, followed closely by Yves Constant Madzou, the young Congolese biologist serving as his scientific apprentice. Farther back, beyond earshot so as not to spook animals, came the larger and noisier group—12 Bambenzele crewmen, bearing heavy loads, and a Bakwele Bantu named Jean Gouomoth, nicknamed Fafa, Fay's all-purpose expedition sergeant and camp cook. They had proceeded that way for many weeks, in a good rhythm, making reasonable distance for reasonable exertion, when gradually they found themselves submerged in a swale of vegetation unlike anything Fay had ever seen.

Trained as a botanist long before he did his doctoral dissertation on gorillas, Fay described it as "a solid sea of Marantaceae"—the family Marantaceae constituting a group of herbaceous tropical plants, including gangly vines such as that thorny nuisance, *Haumania danckelmaniana,* and its near cousin *Haumania liebrechtsiana,* a more vertically inclined plant that can grow into stultifying thickets, denser than sugarcane, denser than grass, dense as the fur on a duck dog. The Marantaceae brake that Fay and his team had now entered, just east of the Sangha River, stretched westward for God only knew how far. Fay himself, with a GPS unit and a half-decent map but no godlike perspective, knew not. All he could do was point Mambeleme into the stuff, like a human Weedwacker, and fall in behind.

Sometimes they moved only 60 steps an hour. During one 10-hour day, they made less than a mile. The green stems stood 15 feet high, with multiple branches groping crosswise and upward, big leaves turned greedily toward the sun. "It's an environment which is completely claustrophobic," Fay said later, from the comfort of retrospect. "It's like digging a tunnel except there is sunlight." The cut stems scratched at the bare arms and legs of him and his crew. Sizable trees, offering shade, harboring monkeys, were few. Flowing water was rare, and each afternoon they searched urgently for some drinkable sump beside which to camp. When they did stop, it took an hour of further cutting just to clear space for the tents.

On the march, Fay spent much of his time bent at the waist, crouching through Mambeleme's tunnel. He learned to summon a Zen-like state of self-control, patience, humility. The alternative was to start hating every stem of this Marantaceae hell, regretting he ever blundered into it—and along that route a person might go completely nuts. Mambeleme and the other Bambenzele had their own form of Zen-like accommodation. *"Eyali djama,"* they would say in Lingala. *"Njamba eyaliboyé*—That's the forest. That's the way it is."

But this wasn't the real forest, woody and canopied and diverse, that Mike Fay had set out to explore. It was something else—an awesome expanse of reedy sameness that he later dubbed the Green Abyss.

They reached the Sangha River, crossed in borrowed pirogues, then plunged westward into more of the same stuff. Fay had flown this whole route in his Cessna, scouting it carefully, but even at low elevation he hadn't grasped the difficulty of getting through on foot. Villagers on the Sangha, whose own hunting and fishing explorations had taught them to steer clear of that trackless mess, warned him: "It's impossible. You cannot do it. You will fail. You will be back here soon." Fay's response was: "We have maps. We have a compass." He had technology: GPS, satellite phone, steel machetes. "We will make it." It sounded overconfident, but on the bottom line he was right. They made it. That took 10 miserable weeks. Having spent New Year's Eve in the Green Abyss, he wouldn't emerge until early March.

"We drank swamp water for three weeks in a row. We did not see any flowing water for almost a month," Fay recalled. "Miraculously, we only had one night where we had to drink water out of a mud hole." It was an old termite mound, excavated by an aardvark or some other insectivore and lately filled with rainwater. The water was thick with suspended clay, grayish brown like latte but tasting more like milk of magnesia.

Food was another problem, since their most recent rendezvous with Fay's logistical support man, the ever-reliable Tomo Nishihara, a Japanese ecologist Fay had known for years, was back at the Sangha River crossing in mid-December. They were now days behind schedule and would be on starveling rations long before they reached the next resupply point. So Fay used the satellite phone to call Tomo and arranged an airdrop: 20-kilogram bags of manioc and 50-can cases of sardines dumped without parachutes from a low-flying plane at an agreed point within the Green Abyss, on January 6, 2000. They made visual contact and the drop was a success, despite one parcel ripping open on a tree limb, leaving a plume of powdered manioc to sift down like snow and 50 sardine cans mushed together like a crashed Corvair. Fay and his crew binged on the open sardines, then resumed walking.

Other problems were less easily solved. There were tensions and deep glooms. There were days that passed into weeks not just without flowing water but also without civil conversation. Not everyone on the team found his own variant of *Njamba eyaliboyé*. By the time they reached the Lengoué River in mid-January, almost four months into the expedition, Madzou'd had enough, and Fay'd had enough of his enoughness. By mutual agreement Madzou left the Megatransect to pursue, as the saying goes, other interests. He was human, after all.

Fay was Fay. He marched on.

AFTER SIX MONTHS, Fay and his crew paused for rest and resupply at a field camp called Ekania, on the upper Mambili River, within another spectacular area of Congolese landscape, this one protected as Odzala National Park. Odzala is noted for its big populations of forest elephants and gorillas, which show themselves in the bais, those small meadowy clearings sparsely polka-dotting the forest. Mineral salts, edible sedges, and other nutritious vegetation at a bai attract not just elephants and gorillas but also forest buffalo, sitatungas, bongos, and red river hogs, sometimes in large groups. Of course, Fay wanted to visit the bais, which he had scouted by plane but never explored on foot; he also wanted to take the measure of the forest around them.

Odzala's elephants suffered heavily from poaching during the late 1980s and early '90s, until a conservation program known as ECOFAC, funded by the European Commission, assumed responsibility for managing the park, with a stringent campaign of guard patrols and a guard post on the lower Mambili to choke off the ivory traffic coming downriver. Access deep into Odzala along the Mambili, a chocolaty stream whose upper reaches are narrow and strained by many fallen trees, was still allowed for innocent travelers not carrying tusks. That's how Tomo brought the resupply crates up to Ekania. It was a 10-hour trip by motorized dugout from the nearest grass airstrip, and on this occasion, I traveled with him.

Fay, bare-chested and walnut brown, with a wilder mane of graying hair than I remembered, stood on a thatched veranda near the camp's boat landing, taking video of us as we docked. Without pulling the camera from his eye, he waved. I can't remember if I waved back; more likely I saluted. He had begun to remind me of a half-mad, half-brilliant military commander gone AWOL into wars of his own choosing, with an army of tattered acolytes attending him slavishly, rather like Brando's version of Conrad's Kurtz in *Apocalypse Now,* only much skinnier.

It was the first time I had seen Fay since Day 13 of the Megatransect, back in October, when I split off from his forest trek and walked out to a road. Now his shoulder bones stood up like the knobbled back of a wooden chair, suggesting he'd lost 20 or 30 pounds. But his legs were the legs of a marathoner. The quiet, clinical smile still lurked behind his wire-rim glasses. Greeting him again here on Day 182, many hundreds of miles deep in the equatorial outback, I felt like Stanley addressing Dr. Livingstone.

"Every day that I walk," Fay volunteered, "I'm just happier that I did the Megatransect." He said "did" rather than "am doing," I noticed, though in fact he was only halfway along. Why use past tense? Because the advance planning and selling phase had been the most onerous part, I suspected, after which the actual walk felt like raking in a poker game pot. Aside from a chest cold and a few foot worm infections, and notwithstanding the weight loss, he had stayed healthy. His body seemed to have reached some sort of equilibrium with the rigors of the forest, he said; his feet, I saw, were marked with pinkish scar tissue and pale sandal-strap bands against the weathered brown. No malaria flare-ups, no yellow fever. Just as important, he was having fun—most of the time, anyway. He described his 10 weeks in the Green Abyss, making clear *that* passage, far from fun, had been "the most trying thing I've ever done in my life." But now he was in Odzala, lovely Odzala, where the bongo and the buffalo roam. He had a new field companion to help with the botany, a jovial Congolese man named Gregoire Kossa-Kossa, forest hardy and consummately knowledgeable, on loan from the Ministry of Forestry and Fishing. Fafa, his crew boss and cook, had grown into a larger role, which included data-gathering chores Madzou earlier handled. And his point man, Mambeleme, now with a buffed-out right arm and a machete so often sharpened that it was almost used up, had proven himself a champion among trail cutters. The rest of Fay's crew, including the brothers Kati and Mouko, had suffered badly from that chest cold they all caught during a village stop, but now seemed fine. Mouko's more serious illness, along with other tribulations, was yet to come.

Meanwhile, Fay's own data gathering had continued, providing some new and significant impressions of Odzala National Park. For instance, one day in a remote floodplain forest, he, along with Mambeleme and Kossa-Kossa, had sighted a black colobus monkey *(Colobus satanas)*, the first report of that rare creature within the park. In the famed bais of Odzala, he saw plenty of elephants, as he had expected, but during his long cross-country traverses between one bai and another, he found a notable absence of elephant trails and dung, suggesting that a person shouldn't extrapolate from those bais to an assumption of overall elephant abundance. His elephant-sign tallies, recorded methodically in the yellow notebook, would complement observations of elephant distribution made by ECOFAC researchers themselves.

Maybe those notebooks would yield other insights too. Maybe the Megatransect wasn't just an athletic publicity stunt, as his critics had claimed. It occurred to me as an intriguing possibility, not for the first time, that maybe Mike Fay wasn't as crazy as he looked.

After a few days at Ekania we set off toward the Mambili headwaters and a large bai called Maya North, near another ECOFAC field camp used by elephant researchers and visiting film crews. The usual route to Maya North camp was upriver along the Mambili, traveling some hours by motorized dugout to a point where ECOFAC workers had cut a good trail. We came the back way, bushwhacking on an overland diagonal. That evening, as we sat by the campfire trading chitchat with several Congolese camp workers, the talk turned to boat travel on the upper Mambili. Well, we didn't use a boat, Fay mentioned. You didn't? they wondered. Then how did you get here? We walked, Fay said. Walked? All the way from Ekania? There's no trail! True but irrelevant, Fay said.

At daybreak on Day 188, we were at the bai, watching 18 elephants in the fresh light of dawn as they drank and groped for minerals in the stream. Some distance from the others stood an ancient female, emaciated, failing, her skull and pelvic bones draped starkly with slack gray skin. Amid the herd was a massive bull who swept his raised trunk back and forth like a periscope, tasting the air vigilantly for unwelcome scents. He caught ours. There was a subtle shift in mood, then the bull initiated a deliberate, wary leave-taking. One elephant after another waded off toward the far side of the bai, disappearing into the trees. By sunup, they were gone.

By midday, so were we, walking on.

FROM THE UPPER Mambili, Fay planned to ascend toward an escarpment that forms the divide between the Congo River Basin and a lesser system, the Ogooué River Basin, which drains to the Atlantic through Gabon. I would peel off again on Day 195, using another resupply rendezvous with Tomo as my chance to exit. As it happened, Tomo needed three boatmen and a chainsaw to get his load of supplies that far up the snag-choked Mambili, but going back downriver would be easier, and we figured to reach the airstrip in two days.

On the morning of the day of my departure, Fafa was laid flat by a malarial fever, so Fay himself oversaw the sorting and packing of new supplies: sacks of manioc and rice and sugar, cans of peanut butter and sardines, bundles of salted fish, big plastic canisters of pepper and dried onions, cooking oil, granola bars, freeze-dried meats, cigarettes for the crew, many

AA batteries, a fresh stack of colorful plastic bowls, and one package of seaweed, recommended by Tomo as a complement to the salted fish. Finally the packs were ready, the tents struck; Fafa rallied from his fever, and I walked along behind Fay and Mambeleme into the early afternoon.

Fay and I had agreed where I would rejoin him next: at an extraordinary set of granite domes, known as inselbergs ("island mountains"), that rise like huge stony gumdrops from a forest in northeastern Gabon. The forest, called Minkébé, was ecologically rich but microbially menacing; it harbored the infamous Ebola virus. Many months earlier, as we had knelt over a large map (a composite from good national maps published by a French geographical institute, one for ROC and one for Gabon, assembled by me with scissors and tape to accommodate Fay's entire route) on the floor of an office at the National Geographic Society in Washington, D.C., the Minkébé forest was where Fay had written "Ebola region" in red ink. "We'll meet you on the other side of the continental divide," he told me cheerily now. "On our way to the Atlantic Ocean."

Backtracking on the trail to catch Tomo's boat, I shook hands with Kossa-Kossa, Fafa, and each of the crew, thanking them for their good company and support. I was fascinated by these rough-and-ready Bambenzele, whom Fay had somehow cajoled and bullied across hundreds of miles, leading them so far from their home forest into an alien landscape, an alien realm of experiences. They had been challenged beyond imagining, stressed fearfully, but so far they hadn't broken; they put me in mind of the sort of Portuguese seamen, uneducated, trusting, adaptable, who must have sailed with Ferdinand Magellan. By way of farewell, I told them in bad Lingala: *"Na kotala yo, na sanza mibalé*—I'll see you in two months."

I was wrong. It would be three months before Fay reached the inselbergs, an interval encompassing some of his most hellish times since the Green Abyss. And when I did rejoin him there, Mambeleme and all the others would be gone.

FAY AND HIS team followed the escarpment northward along its crest, a great uplifted rim that may have once marked the bank of an ancient body of water. Kossa-Kossa left the troop, as planned, to return to his real-life duties. The others shifted direction again, heading into a thumb of territory where the Republic of the Congo obtrudes westward against Gabon. They struck toward the Ouaga River and found it defended by a huge swamp, which at first seemed passable but grew uglier as they committed themselves deeper. By insidious degrees, it became a nightmare of raffia palms and giant pandanus standing in four feet of black water and mud, the long pandanus leaves armed with rows of what Fay recalled as "horrid cat-claw spines." He and the crew spent two nights there in a small cluster of trees, among which they built elevated log platforms to hold their tents above the muck. Pushing forward, Fay saw the route get worse: deeper water, no trees, only more raffia and cat-claw pandanus, and five days' distance of such slogging still ahead, with a chance that any rainstorm would raise the water and trap them. Finally he ordered retreat, a rare thing for Fay, and resigned himself to a long detour through a zone for which he had no map.

After circumventing the Ouaga swamp, they converged with a human trail, a simple forest footpath that served as an important highway linking villages in that northwestern Congo thumb. The footpath brought them to a village called Poumba, where they picked up two pieces of bad news: that the Gabonese border crossing would be difficult at best, due to festering discord between local authorities on the two sides, and that the ivory trader I've mentioned, known to carry gold, had vanished along that very footpath under circumstances suggesting foul play. From a certain perspective (one that the local gendarmerie might well embrace), the trader's disappearance coincided suspiciously with another bit of odd news: that a white man with an entourage of Bambenzele had materialized from the forest on a transcontinental stroll to count aardvark burrows and elephant dung (so he claimed) and was making fast tracks for the Gabonese border. It could look very incriminating, Fay knew. He felt both eager to move and reluctant to seem panicky. Added to those concerns was another, seemingly minor. For the third time in two weeks one of the Bambenzele, Mouko this time, seemed to be suffering malaria. But a dose of Quinimax would fix that, Fay thought.

Over the next few days, Mouko got weaker. He couldn't lug his pack. At times he couldn't even walk and had to be carried. Evidently it was hepatitis, not malaria, because his urine was dark, the Quinimax brought no improvement, and his eyes were going yellow. Fay slowed the pace and took a turn carrying Mouko's pack. Hiding his uncertainty, he wondered what to do. *All the Pygmies think Mouko is going to die now,* he wrote in his notebook on Day 241. Mouko seemed languid as well as sick, with little will to live, while the others had already turned fatalistic about his death. Fay himself became Mouko's chief nurse. He scolded the crew against sharing Mouko's manioc, using his plate, making cuts on his back to bleed him, and various other careless or well-meant practices that could spread the infection. To the notebook, Fay confided: *I am so sick and tired of being the parent of 13 children, it is too much. Thank god I never had children—way too much of a burden. Solo is the way to go—depend on yourself only. The trouble in a group like this is it's like you're an organism. If one part of you is sick or lost the whole organism suffers.* For another 10 days after that entry, Mouko's survival remained in doubt.

They pushed toward Garabinzam, a village near the west end of the footpath, on a navigable tributary of the Ivindo River, which drains into Gabon. On the last day of walking to Garabinzam, the team covered nine miles, Kati carrying his brother Mouko piggyback for most of the way. That evening, Fay wrote: *I need to ship these boys home. You can just tell they are haggard, totally worn out. No matter how good they were they are just going to go down one by one. I would love to keep my friends but I would be betraying them if I made them stay on any longer—it would be unjust.*

Several days later, he departed from his line of march—and from all his resolutions about continuity—to evacuate Mouko downriver by boat. They would try for a village at the Ivindo confluence, on the Gabonese side; from there, if Mouko survived, he could be moved to a hospital in the town of Makokou, about 60 miles south, a hub along one of the few roads in eastern Gabon. Fafa would meanwhile escort the others back to their home forest, hundreds of miles east, sparing them from the onward trudge and the unwelcoming border. Fay himself would pick up the hike in Gabon. One stretch of the planned route would remain unwalked—roughly 18 miles, from Garabinzam overland to the border—as a rankling gap in the data set, a blemish on the grand enterprise, and a token (this was my view, not his) of Fay's humanity.

Left Garabinzam, all is well, he wrote briskly on May 24, 2000, which in Megatransect numeration was Day 248. But he also wrote, almost plaintively: *Pygmies didn't say goodbye.*

MOUKO SURVIVED AND went home. Starting from scratch, Fay gathered a new crew from the villages and gold-mining camps of the upper Ivindo River region there in northeastern Gabon. He found an able young Aka named Bebe, with good ears for wildlife and a strong machete arm, who emerged before long as his new point man; he found a new cook and eight other forest-tough Aka and Bantu men; he found energy, even enthusiasm, to continue. They set off on a long arc through the Minkébé forest, targeting various points of interest, most dramatic of which were the inselbergs. That's where I next saw Fay, on Day 292, when Tomo and I stepped out of a chartered helicopter that had landed precariously on one of the smaller mounds.

Skin browner, hair longer and whiter, Fay looked otherwise unchanged. Same pair of river shorts, same sandals, same dry little smile. I had brought him three pounds of freshly ground coffee and a copy of Michael Herr's 1977 book *Dispatches,* a memoir of the Vietnam War, because I knew he found that genre fascinating. If he was pleased to see me, for the company, for the coffee, he gave no sign.

At once he began talking about data. He had been spotting some interesting trends—for instance, regarding the gorillas. It's true, he said—picking up a discussion from months earlier—that there's a notable absence of gorillas in the Minkébé forest. Since crossing the border, he hadn't heard a single chest-beat display and he had seen only one pile of gorilla dung. Back in Odzala National Park, over a similar stretch, he would have counted three or four *hundred* gorilla dung piles. Elephants were abundant; duikers and monkeys and pigs, abundant. But the gorillas were missing. He suspected they had been wiped out by Ebola.

The Minkébé forest block, encompassing more than 12,500 square miles of northeastern Gabon, represented one of the great zones of deep wilderness remaining in Central Africa. Much of it stood threatened by logging operations, bush meat extraction such as inevitably accompanies logging, and elephant poaching for ivory. But the Gabonese government had recently taken the laudable step of designating a sizable fraction (2,169 square miles) of that block as the Minkébé Reserve, a protected area; in addition, three large adjacent parcels were being considered for possible inclusion. The Gabonese Ministry of Water and Forests, with technical help and gentle coaxing from the World Wildlife Fund, had been studying the farsighted idea that an enlarged Minkébé Reserve might be valuable not just in ecological terms but also in economic ones for its role in the sequestration of carbon. With greenhouse gases and climate change becoming ever more conspicuous as a global concern (this was now July 2000), maybe other nations and interested parties might soon be willing to compensate Gabon—so went the logic—for maintaining vast, uncombusted carbon storehouses such as Minkébé.

But before the reserve extension could be approved, on-the-ground assessments were requisite. So in the past several years a small group of scientists and forest workers had made reconnaissance expeditions into Minkébé—both the original reserve and the proposed extension. They found spectacular zones of forest and swamp, stunning inselbergs, networks of streams, all rich with biological diversity and virtually untouched by human presence. They also found—as Mike Fay was now finding—a nearly total absence of gorillas and chimpanzees.

It wasn't always so. In a 1984 paper in the *American Journal of Primatology,* by Caroline Tutin and Michel Fernandez, the authors described their census of gorilla and chimpanzee populations throughout Gabon. Using a combination of field transects, habitat analysis, and cautious extrapolation, Tutin and Fernandez estimated that at least 4,171 gorillas lived within the Minkébé sector, representing a modest but significant population density. Something seemed to have happened to those apes between 1984 and 2000.

It may have happened abruptly in the mid-1990s, when three Ebola outbreaks burned through villages and gold camps at the Minkébé periphery, killing dozens of humans. One of those events occurred in early 1996 at a village called Mayibout 2, on the upper Ivindo River. It began with a chimpanzee carcass, killed by hunters or found dead in the forest (and perhaps claimed as a kill), then brought to the village as food. Eighteen people who helped with skinning, butchering, and handling the chimp flesh became sick. Suffering variously from fever, headache, and bloody diarrhea, they were evacuated downriver to the hospital at Makokou, the regional hub town. Four of them died quickly. A fifth escaped from the hospital, went back to Mayibout 2, and died there. That victim was buried in the traditional way—ceremonies were performed, and no special precautions were taken against infection.

This bare record of facts and numbers came from a report published three years later, by Dr. Alain-Jean Georges and a long list of co-authors, in a special supplement to *The Journal of Infectious Diseases*. Although the raw chimp flesh had been infectious, the cooked meat evidently hadn't been; no one got sick, the Georges paper asserted, simply from eating it. But once the outbreak began, there were some secondary cases, one human victim infecting another, and the disease spread from Mayibout 2 to a couple of villages nearby, Mayibout 1 and Mvadi. By early March, 31 people had fallen ill, of whom 21 died, for a mortality rate of almost 68 percent. Then it was over, as abruptly as it started. Around the same time, according to later accounts, dead gorillas were seen in the forest.

Mike Fay wasn't the only person inclined to connect Minkébé's shortage of gorillas with Ebola virus. Down in the Gabonese capital, Libreville, I heard the same idea from a lanky Dutchman named Bas Huijbregts, associated with the World Wildlife Fund's Minkébé Project, who had made some of those reconnaissance hikes through the Minkébé forest, gathering both quantified field data and anecdotal testimony. Gorilla nests, Huijbregts reported, were drastically less abundant than they had been a decade earlier. About the gorillas themselves, he said: "If you talk to all the fishermen, hunters, gold miners, they all have a similar story. Before there were many—and then they started dying off." The apparent population collapse, not just of gorillas but of chimps too, seemed to coincide with the human outbreaks. In a hunting camp just north of the Gabonese border, someone showed Huijbregts the grave of a man who, so it was said, had died after eating flesh from a gorilla he had found dead in the forest.

I spoke also with Sally Lahm, an American ecologist who had worked in the region for almost 20 years. Lahm focused especially on the mining camps of the upper Ivindo, where gold came as precious flecks from buried stream sediments and protein came as bush meat from the forest. Her studies of wildlife and its uses by humans, plus the epidemic events of the mid-1990s, had led her toward the subject of Ebola. When the third outbreak occurred, at a logging camp southwest of Minkébé, she went there with several medical people from the Makokou hospital and played a double role, as both nurse and researcher.

"I'm scared to death of Ebola, because I've seen what it can do," Lahm told me. "I've seen it kill people—up close." Fearful or not, she was engrossed by the scientific questions. Where did Ebola lurk between outbreaks? What kind of creature in the forest—a small mammal? an insect?—served as its "reservoir host," the living sanctum within which it abided inconspicuously, over time, between the dramatic episodes of emergence into people or apes? (A virus, which is a genetic parasite, not a cellular life-form, can only replicate itself within the living cells of its hosts. Hence the need for a reservoir creature in which it exists between human outbreaks.) How did its ecology intersect the ecology of hunters, villagers, miners? So far, nobody knew.

"It's not a purely human disease," Lahm said. "Humans are the last in the chain of events. I think we should be looking at it as a wildlife-human disease." Besides doing systematic field research, she had gathered testimony from hunters, gold miners, survivors of Mayibout 2. She had also made field collections of tissue from a whole range of reservoir-candidate animals, shipping her specimens off to a virology institute in South Africa for analysis. And she had grown suspicious of a certain animal, which she judged a prime suspect as the virus's reservoir host. But she declined to tell me what animal that was. She needed to do further work, she explained, before further talk.

On the evening of Day 299, at Fay's campfire, I heard more on this subject from one of his crewmen, an affable French-speaking Bantu named Thony M'Both. *Mayibout deux? Ah, oui,* yes, he was there; he recalled the catastrophe well. Fay translated where my French was lacking. Yes, it began with the chimpanzee, Thony said. Some boys had gone hunting with their dogs; they were after porcupine, and they found the chimp, already dead. No, they didn't claim they had killed it. The body was rotten, belly swelling, anyone could tell. Many people helped butcher and cook it. Cook it how? In a normal African sauce. All who ate the meat or touched it got sick, according to Thony. Vomiting and diarrhea. Eleven victims were taken downriver to the hospital—only that many, because there wasn't enough fuel to carry everyone. Eighteen stayed in the village, died there, were buried there. Doctors came up from Franceville (in southern Gabon, site of a medical research institute) wearing their white suits and helmets, but so far as Thony could see, they didn't save anyone.

Thony's friend Sophiano lost six family members, including his sister-in-law and three nieces. Sophiano Etouck (another of Fay's crew, also seated at the campfire) held one niece in his arms as she died, Thony said, yet he didn't get sick. Nor did Thony himself. He hadn't partaken of the chimp stew. He doesn't eat chimpanzee or gorilla, Thony averred, implying that it was culinary scruple, not from fear of infection. Nowadays in Mayibout 2, however, *nobody* eats chimpanzee. All the boys who went porcupine hunting that day, they all died, yes. The dogs? No, the dogs didn't die.

The campfire chatter around us had stilled. Sophiano himself, a severe-looking Bantu gold miner with a bodybuilder's physique, a black goatee, a sweet disposition, and an anguished stutter, sat quietly while Thony told the tale.

I asked one final question: Had you ever before seen such a disease? I was mindful of the medical reports and other accounts about horrible, chain-reaction Ebola episodes, with victims bleeding profusely, organ shutdown, chaotic hospital conditions, and desperate efforts to nurse or mop up, leading only to further infection. "No," Thony answered blandly. "This was the first time."

Thony's body count differed from the careful, dry narrative in *The Journal of Infectious Diseases,* and so did some other particulars of his version, yet his eyewitness testimony seemed utterly real. He was as scared of Ebola—this virus, this disease, this sinister visitation, whatever it was—as anybody. If he were inventing, would he invent the chimpanzee's swollen belly? I thought not. Added to it all, though, was one fact or factoid that Thony let drop on the first evening I met him, a detail so garish, so perfectly dramatic, that even having heard it from his lips I was unsure whether to take it literally. Around the same time as the Mayibout epidemic, Thony told me, he and Sophiano saw a whole pile of gorillas, 13 of them, lying dead in the forest.

Anecdotal testimony, even from eyewitnesses, tends to be shimmery, inexact, unreliable. To say *13 dead gorillas* might actually mean a dozen, or a lot—too many for a startled brain to count. To say *I saw them* might mean exactly that or possibly less. *My friend saw them, he's unimpeachable.* Or maybe, *I heard about it on pretty good authority.*

Scientific data are something else. They don't shimmer with poetic hyperbole and ambivalence. They are particulate, quantifiable, firm. Fastidiously gathered, rigorously sorted, they can reveal emergent meanings. That's why Mike Fay was walking across Central Africa with a little yellow notebook.

After two weeks of bushwhacking through Ebola's backyard, we emerged from the forest onto a red laterite road, the highway between Makokou and the world. Blinking against the sunlight, we found ourselves in a village called Minkoula, at which the dependable Tomo soon arrived with more supplies. Day 307 ended with us camped in a banana grove behind the house of a local official, flanked by a garbage dump and a gas-engine generator. The crew was given an evening's furlough, and half of them caught rides into Makokou to chase women and get drunk. By the next morning, one of the Aka would be in jail, having expensively busted up a bar, and Fay would face a new round of political hassles, personnel crises, and minor ransom demands, a category of inescapable chores he found far less agreeable than walking through swamp. But somehow he would get the crew moving again. He would plunge away from the red road, diving back into the universe of green. Meanwhile, he spent hours in his tent, amid the bananas, collating the latest harvest of data on his laptop.

Then he emerged. Within the past 14 days, he informed me, we had stepped across 997 piles of elephant dung and not a single dung pile from a gorilla. We had heard zero gorilla chest-beat displays. We had seen zero sprigs of Marantaceae chewed by gorilla teeth and discarded. These numbers represented a good metric of the mystery of Minkébé. Their implication: The gorillas were all dead of Ebola—or, if any survived, they had fled that malign place. But the reservoir host was still unknown.

Measuring one aspect of that mystery with zeroes was a crucial first step. Solving it was another matter.

I made my departure along the laterite road and then by Cessna from the Makokou airstrip. The pilot who came to chauffeur me was a young Frenchman named Nicolas Kozon, the same fellow who had circled the Green Abyss at low altitude while Tomo tossed bombs of manioc and sardines to Fay and the others below. Now, as we rose from the runway, climbed farther, and pointed ourselves toward Libreville, Gabon's capital, the road and the villages disappeared quickly, leaving Nicolas and me with a limitless vista of green. Below us, around us in all directions to the horizon, there was only forest canopy, and more canopy, magisterial and abstract.

Nicolas was both puzzled and amused by the epic daffiness of the Megatransect, and through our crackly headsets we discussed it. I described the daily routine, the distances made, the swamps crossed, and what Fay faced from here onward. He would visit the big waterfalls of the Ivindo River, I said, then turn westward. (I knew his plotted route well, because he had inked it onto that taped-together map, which I still carried.) He would cross the railroad line and two more roads, but otherwise he'd keep to the forest, following the route, staying as far as possible from human settlements. He could do that all the way to the ocean. He would cross the

Lopé Reserve in central Gabon, yes, and then a big block of little-known terrain around the Massif du Chaillu. Another four months of walking, if all went well. He was skinny, but looked strong. He would cross the Gamba complex of defunct hunting areas and faunal reserves along the coast, south of Port-Gentil, and break out onto the beach. He expected to get there in late November, I said.

With a flicker of smile, Nicolas asked: "And then will he swim to America?"

. . . .

The mystery of Ebola's reservoir host continued to puzzle scientists—it continues still—and Sally Lahm never did tell me which species of forest creature she considered the prime suspect. Was it a monkey? Was it a rodent? Was it a bat?

Some people, nonexperts speaking carelessly, will indeed state on television or elsewhere that Ebola comes from bats, but they can't specify which kind of bat (there are more than 1,400 species) because they don't know and no one else does either. Yes, fragments of Ebola's RNA genome and antibodies against Ebola virus have been detected in some bats, but that could represent evidence only of transient infection, or mere exposure to the virus, not necessarily a long-term role in hosting it. The gold standard for identification of a reservoir host is to culture (grow within cells in a laboratory) "live" virus, intact and functional, from a sample of blood, feces, or other material taken from the reservoir candidate in question. As of this writing, that has never been accomplished with Ebola virus. I say so with confidence, having checked the point just months ago with one of the world's leading Ebola scholars, Jens Kuhn, a virologist of broad experience, a historian of virology, and author of the book Filoviruses: A Compendium of 40 Years of Epidemiological, Clinical, and Laboratory Studies. *Ebola virus, along with Marburg virus and a few other viruses of filamentous structure and (most of them) high virulence in humans, is one of the filoviruses.*

I happen to know Jens Kuhn, as a friend as well as a go-to source, because my experience with Fay in the Minkébé forest, with its ghostly absence of gorillas, changed my professional life in the years after the Megatransect. Looking into the question of where Ebola virus might lurk within Minkébé's native fauna, how it might have spilled over into humans at Mayibout 2, and how it happened to be adaptive enough to flourish within a new kind of host, brought me to realize that the subject of Ebola, and of other scary new viruses, lies within the disciplines of ecology and evolutionary biology.

Those disciplines being my accustomed beat, I found it both natural and urgent to investigate further and write more about novel viruses that emerge from nonhuman animals to infect humans and, sometimes, cause sizable outbreaks, epidemics, and pandemics. The subject struck me as important and urgent, in terms of conservation as well as human health, because the disruption by humans of highly diverse ecosystems (such as tropical forests) is known to be one of the factors that create opportunities for those viruses to spill over from bats, rodents, apes, and other nonhuman animals into people. For instance: Machupo virus (1963), Marburg virus (1967), HIV-1 (the pandemic AIDS strain, recognized as a human pathogen in 1981), Nipah virus (1998), Zika virus (2015), and others.

If you're curious, see my 2012 book Spillover, *the direct result of note-taking and thinking that began as I sat at that campfire in Minkébé with Thony M'Both and Mike Fay and Sophiano Etouck, on July 14, 2000. I still have the notebook. There's a dried brown stain on the cardboard front flap, which to my best recollection represents blood from a mosquito that bit me and got smacked.*

Minkébé itself has changed substantially since that campfire evening, and at least partly for the better. What was then the Minkébé Reserve is now Minkébé National Park, one of 13 new national parks decreed by Gabon's president El Hadj Omar Bongo in 2002, as a result of Fay's Megatransect and other factors. (See "A Country to Discover," pages 81–96.)

*Odzala National Park in the Republic of the Congo has changed too, in several ways, not all good. When I wrote that Fay had found Odzala far richer in signs of gorilla presence than Minkébé—many more chest-beat displays heard, many more piles of gorilla dung spotted—I was referring to Odzala as it had been in March 2000, when Fay (and I with him, for part of that leg) walked through it. But during 2002 and 2003, Ebola struck also in Odzala and round about, killing dozens of people in villages and hundreds—probably thousands—of gorillas in the forest. That's also part of another story in another book (*Spillover *again), only foreshadowed in this one.*

Several years after the Megatransect, Odzala was expanded into Odzala-Kokoua National Park, and since 2010, it has been managed by a private organization, African Parks (see "Boots on the Ground," pages 309–324), under a 25-year agreement with the government of the Republic of the Congo. Notwithstanding the efforts of African Parks (AP), Odzala's gorilla population continued to decline, as a combined result of poaching and Ebola virus, from an estimate of 40,000 animals in 2005, to 22,000 in 2012, to fewer than 8,000, according to AP's wildlife survey of 2020. AP estimates that roughly 12,000 people live around the periphery of Odzala, hungry adults and children, all needing protein, and so poaching wildlife for subsistence (and for sale) remains a big threat. More than 14,000 snares and 55 tons of bush meat were seized there in 2019. And the linkage among the Ebola virus, susceptible gorillas and chimps, and the consumption of ape meat (sometimes from shooting or trapping, sometimes from scavenging) by humans is what makes the situation so dangerous for all concerned.

THE MEGATRANSECT III

-·|·|·|·-

The End of the Line

· ·

At the beginning, and again at some points along the way, it seemed almost impossible that Fay could complete this epic trek as projected, given the factors such as exhaustion, malaria, the politics of national boundaries, the finitude of patience and stamina among the men of his field crew, the occasional charging elephant, the swamp water to drink, the filaria flies to shoo off, and the Gaboon vipers to avoid stepping on—all that and the other considerations arrayed against him. But a year had gone by, and he was still on the trail. Then 14 months, and he had passed from the Congo River drainage into the Ogooué Basin of northeastern Gabon, and crossed the Réserve de Lopé-Okanda with its dry savannas and its great, running troops of mandrills, and then topped over the Massif du Chaillu and into the forested lowlands of Gabon's southwestern corner, where the black waters of countless smaller rivers, swamps, and small lakes flowed and seeped toward the sea. And it became likely that he would indeed reach his destination, the Atlantic Ocean.

His will was implacable, his body was small and skinny, but it was toughened, as was his mind, by so many hard miles and stressful days. He was fiercely committed—not just to the walk, the full distance, the achievement, but most of all to the data set he was producing, the tens of thousands of observations in his notebooks, the GPS points, photographs, metrics of all kinds. Clearly, he would finish or die trying, and at this point it seemed unlikely he would die. So, of course, Nick and I wanted to be there to see him finish.

· ·

On the 453rd day of his punishing, obsessional, 15-month hike across the forests of Central Africa, J. Michael Fay stood on the east bank of a body of water, gazing west. It was not the Atlantic Ocean. That goal, the seacoast of southwestern Gabon, the finish line to his trek, was still 20 miles away. And now his path was blocked by a final obstruction, not the most daunting he had faced but nonetheless serious: this large black-water pool, a zone of intermittently flooded forest converted to a finger lake by seasonal rains. Soaked leaf litter and other detritus had yielded the usual tannin-rich opacity, and so the water's sleek surface was as dark as buffed ebony, punctuated sparsely by large trees, their roots and buttresses submerged. Submerged how deeply? Fay didn't know. Eighty yards out, the flooded forest gave way to a flooded thicket, a tangle of dense, scrubby vegetation with low branches and prop roots interlaced like mangroves, forming a barrier to vision and, maybe, to any imaginable mode of human passage. How far through the thicket to dry land? That, also, Fay didn't know.

"This is the moment of truth, I think," he said.

If it's only waist deep, I said, with vapid good cheer, we could easily wade across.

"If it stays no deeper than *shoulder*," he corrected me, "we can make it." But he wasn't optimistic.

Fay took the machete of his point man, Emile Bebe, the young Aka who had cut trail for him across hundreds of miles of Gabon. Slipping off his pack, wearing only his usual amphibious outfit (river sandals and river shorts; he had long since discarded the T-shirt), Fay waded out alone, probing the dark water ahead of him with a long stick. Bebe and I and two other walking companions—photographer Nick Nichols and a videographer from National Geographic Television, Phil Allen—stood watching him go. Soon he was waist-deep, chest-deep, then armpit-deep, groping with his feet against sudden drops, seeking the shallowest route. Then there was just a little head and two skinny arms vanishing into the thicket. I climbed onto a woody loop of liana against the base of a tree, putting me six feet above the water and better positioned to listen, if not to see. I was concerned for him out there alone because of the crocodiles—not just the fearsome Nile croc *(Crocodylus niloticus)*, but also a smaller croc found hereabouts, commonly known as the dwarf crocodile *(Osteolaemus tetraspis)*, not to be taken lightly. Of course my concern was futile, I realized, because from this distance, perched like a parrot on a trapeze, I couldn't give any timely help if a crocodile did grab him. I heard the whack of the machete. I heard fits of cursing, which alternated oddly with what sounded like bursts of semi-demented song. We waited. He was gone for a half hour, 40 minutes, longer.

Meanwhile, the rest of the traveling crew—two other Aka and seven Bantu men, all carrying heavy packs of camp gear and scientific equipment and food, plus a middle-aged Gabonese forestry technician named Augustin Moungazi, whose role was to census trees—caught up and joined us at the water's edge. Where's the boss? they asked. Somewhere out there. The crewmen cast their eyes across this forbidding obstacle—which, in my mind and notebook became the Black Lake, preeminent to all the other little black-water ponds and streams—with varying gradations of weariness and dread. Most of them had worked with Fay seven months now, since he had crossed into Gabon from the Republic of the Congo, and they had been through such moments before. In the way they shrugged off their packs, uncricked their shoulders, inspected the route forward with leery scowls, they seemed to be saying, Oy, what manner of muddling travail gets us around *this* obstacle? It looked bad, but they had seen worse.

After nearly an hour, I climbed down from my perch. Bebe smoked another cigarette. Nick aimed his Leica at anything remotely interesting. We swatted at filaria flies. We ate our crackers, nuts, and other piddling snacks representing lunch. We wondered silently whether Fay would ever come back and, if not, how we'd find our way out of this forest without our mad leader. Then we heard shouts.

He had reached landfall beyond the thicket and returned just far enough to holler instructions. Mainly he was calling to the crewmen, in French, through the wattle of vegetation and the heavy equatorial air.

Admittedly my French is lousy, Nick's and Phil's were even worse, so we were befuddled, yet the francophone crewmen appeared befuddled too. If we could just understand what Mike was saying, all of us, we would gladly comply. But to my ears, he sounded like a bilious colonel of the French Foreign Legion screaming orders at new recruits through a mattress.

He had been right, in some sense, when he called it a moment of truth. Whereas Fay had come to study the forest, I had come to study Fay, and adversity is a great illuminator of true character. But then again, *truth?*—it's a quicksilver commodity, not so easily gathered as data. The moment was still unfolding, and so far there was more confusion than illumination. Did he want us to come or to stay? If we should come, then *how?* Should we cut logs and build a raft, or just swim for it? The voice from the thicket seemed to convey almost nothing but purblind certainty and impatience. Was he mustering his troops for a final heroic lurch? Or, stressed by the long months of walking and the burden of forcing discipline on a group of freely hired men, by the nearness of the end, by his own ambivalence about reaching it, was he having a meltdown?

Days after this episode in the Black Lake, I would still be asking myself those questions. I would still be puzzling over the matter of J. Michael Fay and the complicated, provocative subject of leadership.

It was both the logic and the momentum of Fay's grand enterprise, which he had labeled the Megatransect, that had brought him and his entourage to this point of exigency on the 453rd morning. Those factors hadn't changed since the beginning. The logic was that he would walk a zigzag route from the northeastern corner of the Republic of the Congo to the southwestern coast of Gabon, a distance of maybe 1,200 miles (but much longer once he added all the minor diversions and detours), passing dead center through vast blocks of roadless and uninhabited forest, gathering data on vegetation, wildlife, and forest conditions as he went. The forest blocks, lying contiguous to one another, could be seen as gobbets of raw meat on Africa's last great kebab of tropical wilderness. Fay's route was to be the skewer.

The momentum derived from 452 days of footslog persistence, including many swamps mucked across and creeks forded, many resupply problems, many hungry nights, many nervous elephants with half a notion to make Fay himself a kebab, many hours of campfire laughter and bonhomie with the crew, many explosions of anger, many points at which it seemed almost impossible for Fay and his comrades to go on, after which they went on. Fay's logic insisted that this gargantuan transect be continuous and unbroken, both in space and in time. There had already occurred the one unavoidable gap, back in northwestern ROC just short of the Gabon border, when he departed from his plotted line to evacuate the crewman Mouko, who was verging on death from hepatitis. Although that short unwalked stretch—just 18 miles, which he called the Mouko Gap—continued to nag Fay with a slight sense of incompleteness, he had put it behind him, marching on. By now his momentum included so many miles traversed (more like 2,000 than the 1,200 originally foreseen), and so many crises passed, that it was unthinkable to be balked again, this time within 20 miles of the beach.

The logic of the enterprise had been laid out to the National Geographic Society (his main sponsor) and the Wildlife Conservation Society (his employer) in a 48-page prospectus, with the forest blocks and his route sketched onto a multicolored map. The blocks as he had delineated them numbered 13, beginning with the Nouabalé-Ndoki block in northeastern ROC and ranging southwestward from there. Last in the chain was the Gamba block, a cluster of faunal reserves and defunct hunting areas along the Atlantic coast that were then being organized by the Gabonese government, with help from the World Wildlife Fund, into a complex of protected areas intended to preserve good habitat for elephants, hippos, dwarf crocodiles, and other sensitive species all the way to the beach.

Each of these blocks abutted another, and each was circumscribed by human impact (a road, a rail line, a string of villages along a river) but—this was the crucial part—virtually free of such impact at its interior. Although some armchair experts found it hard to believe, there *were* still sizable patches of African forest not currently occupied by human beings. Fay's concept was to travel by foot with a small support crew through these forest blocks and to measure in multiple dimensions the relationship between such absence of human impact and the ecological richness of the forest.

He had described this data-gathering mission as a "reconnaissance survey," to distinguish it from the more formalized procedure known as the line-transect survey, wherein a field biologist walks and rewalks a short, straight path through the forest, gathering accretions of standardized data with each passage. Instead of cutting a ruler-straight corridor, Fay had elected to use a "path of least resistance" approach, letting the contours and obstacles of the landscape nudge him this way and that against his general compass bearing, and to make a single 1,200-mile walk instead of, say, a thousand or so one-mile laps up and down a familiar snippet of trail. "The path of least resistance has the advantage of leaving the forest intact after passage, a significantly increased sample size because of increased speed, and considerably reduced observer fatigue," he had written in the prospectus. During my own time on the trail with him, totaling eight weeks divided into four stretches, I sometimes recollected the irony of that phrase, "the path of least resistance." It sounded lazily sybaritic, whereas here we were, clambering through still another tropical briar patch and then waddling across still another floodplain of sucking mud.

Now again on the morning of Day 453, as I squinted toward that thicket across the Black Lake, somewhere amid which Fay was hacking branches and yodeling orders, I had cause to wonder, This is the path of least resistance? Thank God we didn't come the hard way.

LIKE AN UNNERVING omen of things to follow, Day 453 began with leeches. We had spent the night at Leech Pond Camp, thus dubbed by me (I named all the camps in my journal, for mnemonic purposes) when Fay returned from his evening bath and reported that 10 leeches had gotten to him while he was rinsing. Leeches in moderation are no big deal, because they don't hurt and don't fly (although, as I was unaware then, they can indeed carry disease-causing bacteria and viruses). But the leeches that greeted us in the camp pond on the 453rd morning were beyond moderation. They swam up like schools of grunion and hooked their thirsty little maws to our ankles and calves, a half dozen here, a half dozen there, resisting slimily as we tried to pull them off. We had leeches under our sandal straps, leeches between our toes, leeches crawling groin-ward, leeches racing to every open sore. Good grief, what had they lived on before we arrived?

Hopping from foot to foot in the shallows, we de-leeched ourselves while Bebe, also dancing and snatching at his feet between machete strokes, felled a small tree to bridge the pond's deeper trough. Then we tightroped across, de-leeched again on solid ground, and continued on our way.

Within a few minutes we heard monkeys jumping through the canopy. Fay did his usual trick, a whistling imitation of the crowned eagle, which provoked raucous alarm calls (*kaa-ko! kaa-ko!*) from the monkeys, allowing him to identify them: red-capped mangabey *(Cercocebus torquatus torquatus),* locally known (from their call) as the *kako.* He scribbled the exact time and species name into his notebook, then took a five-minute sampling of their vocalizations on digital audio. Earlier he had mentioned that this mangabey, with its unmistakable flaming hairdo, was native only to forests near the Atlantic coast; farther inland, months ago, while crossing ROC and eastern Gabon, he had seen plenty of gray-cheeked mangabeys but none of the red-capped. Now here they were, offering a welcome signal that we had entered the coastal zone.

After an hour of easy walking along elephant trails, we had found ourselves blocked by another dark pond. "Bad news, boys," said Fay. It looked as though the rainy-season waters were still up, he explained, which foreboded that there might be many such fingers of flooded forest between us and the coast. "If that's the case, we ain't gonna get through." But with a little scouting we found a fallen-tree bridge across the deep part, and from there waded to dry land.

At the edge of the water stood another tree, a towering hulk with shaggy bark, a gracefully tilted trunk, and wide-reaching buttresses. Fay's routine called for noting every major tree along the route, so this one went into his little book: *Sacoglottis gabonensis,* 1.5 meters diameter near the base. Loggers generally ignore the species, he had said earlier, because its ropy, twisting trunks don't yield good lumber. The increasing abundance of *Sacoglottis gabonensis* was a further indicator that we were nearing the ocean. Still another was *Tieghemella africana,* a tree of high value both to timber companies and elephants. Known commercially as douka, it grows to magisterial sizes—six feet in diameter and crowning out through the canopy—with straight, clean trunks offering lovely wood for the sawmill. It also produces big green fruits, globular and heavy, each filled with sweet-smelling, pumpkiny orange pulp, which at Fay's invitation we sampled—not bad but a little chalky, to my taste. Elephants travel considerable distances to scarf douka fruits when they are ripe and falling, and the well-worn elephant trails we had been following seemed to run like traplines from one douka tree to another. Take away those mature, fruiting trees (by selective logging, for instance) and the local elephant population would lose part of its seasonal diet. But for now, the grand old doukas were still here, showing evidence of recent attention from elephants (fresh dung, gnaw marks in the bark), and so were the elephant trails. We hit another short stretch of good walking, then heard another group of monkeys.

This time, in response to the eagle whistle, there came a low, grunting chortle: *chooga-chooga-chooga-chooga-chooga.* Having heard it many times over the months, even I could recognize that as the alarm call of the gray-cheeked mangabey *(Lophocebus albigena),* another monkey species dependent on fruiting trees. "It looks like the old gray-cheeks are gonna make it to the beach after all," Fay said. "That's cool. I was a little worried, 'cause we hadn't seen them for three or four days." The presence of gray-cheeked mangabey, overlapping here with its red-capped cousin, became another notebook entry. Then again we walked—westward, toward the beach—but only for five minutes, until the Black Lake stopped us cold.

The Black Lake: too wide to bridge and too long to bypass. According to Fay's map, it led northward into the Rembo Ngové floodplain, a riverine morass we didn't care to enter. It stretched southward too, so Fay had gone straight across, on his solitary probe, and was now out there somewhere in the thicket, shouting back instructions. Jean-Paul Ango, one of the youngest and strongest of the crewmen, took his machete to a modest-size tree, which fell pointlessly into the water near shore. That can't be the idea, I thought.

Impatient with this muddle, I waded out along Fay's route to see if I could find the shallowly submerged ridge on which he seemed to have walked. Quickly I was neck-deep. So I decided to swim. Another crewman, Thony M'Both, the man who told tales of Ebola at Mayibout 2, took the same notion at the same time, and we breaststroked across the black water on converging lines toward the thicket. Soon most of the crew followed, some confidently, some reluctant to swim but more reluctant to be left behind. Strung out like a line of ducklings, they floundered variously with their waterproof packs—some of those packs filled with rice and other perishable food, some with expensive and precious electronic gear—which were buoyant but too cumbersome to serve as water wings. Reaching the face of the thicket, Thony and I stopped. We treaded water. There seemed nowhere to go. I climbed up into the buttresses of a half-drowned tree, and one by one the others did likewise. In a neighboring tree I noticed Jacques Bosse, a big square-shouldered Bantu that Fay had hired out of one of those gold-digging camps in northeastern Gabon. With a forceful yank, Jacques hoisted up his pack, to the outside of which was tied a large cook pot. He tossed back his head and muttered disgustedly to the sky that this was *no* kind of work for a man. We were stuck there, treed and frazzled like cats in a Mississippi flood, when Fay came out of the thicket and resumed command.

His first act was to holler sternly at Emmanuel Yeye, the shyest of the Aka, for letting his pack soak in the water rather than pulling it up. This gave way to a scathing harangue against the whole crew. Fay derided them for their fecklessness, their incompetence, their childishness and stupidity and insubordination. It was all in French, but what I missed in vocabulary I could gather from tone. It went on and on.

Nick and I had each witnessed earlier episodes of such castigation, going back to the first days of the Megatransect and Fay's Congo crew of Bambenzele men. We had seen it after the walk through Minkébé, when some of the current crewmen got drunk and disorderly during their furlough at a resupply stop. We had seen it elsewhere. I had even begun to expect it (in my notes I called it, for shorthand, the Riot Act) as a calculated, self-conscious performance that Fay used periodically to restore discipline and focus. But this time both Nick and I felt he was going too far. Fay said blistering things of the sort that only a drill sergeant, an abusive father, or an especially caustic fifth-grade teacher might utter. He ranted and scorned. He recited the crew's failings. *"Ça me rend fou,"* he growled repeatedly. "It makes me crazy." Well, maybe so. At that moment, given our circumstances and the brave plunge these men had just taken, I thought that perhaps our brilliantly unorthodox Dr. Fay had indeed gone off his nut.

I was wrong. Later events and conversations with Fay, combined with what I knew of his personality and background, would convince me that this ultimate Riot Act tirade, as we all hung in trees above the Black Lake, was rational and carefully calibrated. Fay was stressed, yes, but still in control. The deeper I scratched him, the more layers of ornery complexity and courageous bluntness I found. He wasn't always likable; sometimes he seemed piteously isolated; sometimes he seemed cynical and mechanistic about human relations; sometimes, just too demanding and harsh. But in my final judgment, reached slowly, Fay is a formidable man with a strong sense both of mission and of fairness.

"Chaos breaks out very quickly and very easily," he would tell me days afterward, in the quiet of a tent pitched on a sandy hillock overlooking the Atlantic surf. "You've got to be a complete and utter hard-ass. And I don't enjoy being a hard-ass. I do not have some kind of sadistic element in my mind that makes me enjoy dominating people. But if you accepted that responsibility . . ." He paused, thinking back over his 15 months of risky travail, it seemed; then he dropped the second-person pronoun and spoke plainly. "Everything was my responsibility. Anyone who died on the Megatransect, it would have been my responsibility."

Mouko had nearly died of hepatitis, and it was Fay who nursed him until evacuation was possible. A crewman named Roger had almost drowned, tangled in his pack straps at a river crossing, when he larkishly flouted Fay's instructions. There had been several other close calls in water, and several other medical emergencies. "I take that very seriously," Fay said. "And I take the data collection very seriously." All along, he explained, he'd had three overriding goals: to finish the entire walk as originally conceived, to maintain an unbroken regimen of data gathering, and to get everyone through the experience alive. Democracy on the trail and his own popularity on any given day were not even secondary concerns.

The data would eventually be collated, cross-referenced, elaborately crunched and analyzed during the months of his follow-up work. Which forests seem to be richest in gorillas? How quickly do elephants recolonize an area where elephants have been poached? What's the linkage between logging roads and the presence or absence of duikers, forest hogs, cercopithecine monkeys? Everywhere, every which way, he wanted to ask and to answer: What are the correlations? He hoped that question would lead to another: What are the implications for wise management? Fay would write a report or a book, maybe both, and then also make it all available through a website.

"On this website people are going to see very clear patterns," he vowed, as we sat above the beach. "Nobody is going to be able to deny that there is something there." Referencing one slice of data to another would in some cases yield high statistical correlation, and observers (so he imagined, supplying their words) would say, "Wow! Look at this, man. Douka, elephant: correlation, point nine." He meant that the degree of relatedness between the presence of douka trees and the presence of elephant trails might be 0.9, of a possible 1.0. "That's pretty cool." Fay hoped, anyway, that observers would find such a relationship and offer such a response. "Just seeing those patterns is going to make people realize that this is a viable methodology."

But before any such epiphanies could happen, he needed the data. It had to be continuous. It had to be rigorous. Toward that end, his organizational model for the Megatransect was unabashedly autocratic. During one milder fit of annoyance, provoked by a food shortage after some crewmen had evidently jettisoned provisions to lighten their packs, I heard Fay tell the men that if this were a military operation, they would all be in prison. "It's very much like a military operation," he said to me now. "I am the commander in chief on the Megatransect." That might sound "radical" to some ears, he acknowledged (or maybe "offensive" or "autocratic" are the words you want there, I thought), but to anyone who had shared the many months of daily effort and frequent peril, it would make perfect sense. There could be only one leader giving orders, and those orders had to be followed without malingering or debate, or else the whole effort would unravel and the three goals wouldn't be met.

WHERE DID THIS military style come from? Fay himself is too young to have experienced Vietnam or the draft, too old to have signed on for the first Iraq war, and has never served in any branch of the armed forces. It's hard to imagine how he ever could have. Three or four months of basic training and regimentation would no doubt have aggravated his own insubordinate tendencies to a point of court-martial or discharge. But the lore of certain military operations intrigued him—Vietnam particularly, maybe because it was a jungle war and he's a jungle guy. The American fiasco there, I once heard him argue, reflected the plain fact that American troops weren't at home in the ecosystem. (Undoubtedly that had been part of it, along with several millennia worth of cultural differences and political history, plus a few other factors.) During my earlier visits to the line of march, I had brought him some of the better Vietnam memoirs for trail reading—not just Michael Herr's *Dispatches* but also Doug Peacock's *Grizzly Years*, the memoir of a Green Beret medic who came back from the war angry and hurt, and coped with his trauma by devoting himself to close observation of grizzly bears—which seemed to engross Fay for a few hours at night after his data-entry chores. He mentioned that if he weren't an ecologist, he might be tempted to find work as a war photographer. He occasionally asked me (because he knew Doug was a friend of mine) what "Old Peacock" might have thought or done about this or that. And he was fascinated by the Lewis and Clark expedition, which besides being an exploratory trek was a mission, under full discipline, of the United States Army.

Back at the start of the Megatransect, in the disheveled little library of his cabin in Bomassa, the research camp in northeastern ROC that had served Fay in recent years as a base, I had found his own dog-eared and heavily marked copy of *Undaunted Courage,* Stephen Ambrose's account of the life and character of Meriwether Lewis, as revealed most gloriously during his journey with Clark. That journey was, of course, America's own first and greatest Megatransect. One passage in the Ambrose book, completely underlined by Fay, caught my eye: "Two years of study under Thomas Jefferson, followed by his crash course in Philadelphia, had made Lewis into exactly what Jefferson had hoped for in an explorer—a botanist with a good sense of what was known and what was unknown [and] a working vocabulary for description of flora and fauna, a mapmaker who could use celestial instruments properly, a scientist with keen powers of observation, all combined in a woodsman and an officer who could lead a party to the Pacific." A botanist, a woodsman, a leader. Reading that, Fay must have felt some tingle of identification.

Never mind the sad fact that Meriwether Lewis, addled by acclaim and alcohol after his big success, eventually killed himself. Fortunately for Fay, the parallel between him and Lewis wasn't so close. Lewis stepped into a mission that had been dreamed up by President Jefferson, whereas Fay himself, no one else, concocted this one. Lewis and Clark's enterprise was premised upon the goals of commercial exploitation and easy travel for traders, whereas Fay's Megatransect had a drastically different goal: protecting big areas of rich forest from reductive human impact. Fay had a better scientific education than Meriwether Lewis, and unlike Lewis he seemed not very susceptible to booze or self-doubt. Another advantage was that, whereas Lewis headed off into a difficult sort of landscape he'd never seen before, Fay had 20 years' experience in various Central African forests.

He knew the ecosystems from bottom to top—from the plants to the elephants and gorillas. Equally important, he knew how to walk through this world. Beginning in the late 1980s, when he did his doctoral fieldwork on western lowland gorillas in the Central African Republic, tracking them through the forest with his mentor, a local man named Mbutu Clement,

Fay developed the habit of making long, restless explorations by foot. The little Suzuki trail bike that he had used during his Peace Corps days, up in the savanna country near the border with Chad, was no longer useful and no longer necessary. He discovered that by adapting his body and his outfit (river sandals, one pair of shorts, and no shirt, because bare skin is more easily washed and dried than clothing) to local conditions, he could cross flooded forests, streams, boggy clearings, and swamps that most other people considered impassable. He also learned he could walk into a village or town anywhere in Central Africa and, within a day or two, hire a crew of men who were glad for the work of carrying bags and making camps. Employment was scarce, and he paid better than most. He learned how many men were required for transporting this much scientific equipment, that many tents, and enough food to sustain them all for, say, 20 or 25 days between points of resupply. By trial and error, he developed a style of personnel management that worked.

One element of that style was his imperious sense of command. Another was that he never asked anyone to accept discomfort or risk that he wouldn't accept himself. The Greek historian Plutarch, in his life of the Roman general Marius, written 2,000 years ago, declared that "there is nothing a Roman soldier enjoys more than the sight of his commanding officer openly eating the same bread as him, or lying on a plain straw mattress, or lending a hand to dig a ditch or raise a palisade. What they admire in a leader is the willingness to share their danger and hardship, rather than the ability to win them honor and wealth, and they are more fond of officers who are prepared to make efforts alongside them than they are of those who let them take things easy." This is also what made young Major George Washington a successful mounted commander on the Ohio frontier. In Fay's case, it was manioc and salted fish, not Roman bread or Virginia rice; a roll-out pad on the forest floor, not a straw mattress or a rough woolen blanket; and a machete-cut corridor through a black-water thicket, in lieu of a raised palisade or a primitive log stockade.

When I asked Fay later about his blowup at the Black Lake, he conceded that "it certainly looked like I was pissed off, there's no doubt about it."

And yet he hadn't been, he said. It was just another bit of tactical histrionics. From his perspective (though he was too discreet to say so), I had exacerbated the confusion myself when Thony and I triggered the group swim. He had intended to proceed methodically, but my impatience disrupted that. "I was simply taking chaos and putting order into it. And the only way to do that is to say, at the top of your lungs, 'Everybody *stop!* Everyone who is here present, *stop!* Do not move. Do not breathe. Stop. And I'm going to tell you what to do.'"

Fair enough. Though, in the moment, I hadn't waited to be told. I swam back to the east side of the lake, found my own waterproof pack where I had left it, double-checked its seal for the sake of my notebook and binoculars, and swam out again to the thicket. By the time I got there, nudging the pack ahead of me like a water polo ball, the others had begun moving down Fay's hacked-out corridor. The water here seemed to be eight or 10 feet deep. I fell in behind Sophiano Etouck, the most stalwart of the crewmen, and Nick, who was managing somehow to dog-paddle along one-handed, with his pack on his back and his other hand holding the little Leica to his face like a snorkel mask. God love him, Nick even now was shooting. Sophiano led the way, swimming with his right arm and wielding a machete with his left. Every few yards he rose high in the water to whack a limb out of our path, then sank away beneath a boil of bubbles. When Sophiano first went under, and stayed under, Nick and I both worried that he had tangled himself in some vegetation; then, exuberant as an otter, strong as a dolphin, he exploded back up to take another swing with the machete. I followed him for 50 yards through this watery tunnel of limbs and roots, a passable route that Fay had partly opened during his missing hour. Finally the thicket cleared, the water shallowed suddenly, and we climbed up a high bank onto firm ground.

WHILE NICK AND Phil examined their cameras for damage and their bodies for leeches, I dropped my pack and went back in the water to see if I could help with another load. After swimming down one blind alley, I found the tunnel again and retraced it to the east edge of the thicket. Fay was there,

THE HEARTBEAT OF THE WILD

still perched in a tree, having meanwhile swum the lake to retrieve his own pack.

Now he was shepherding along the last of the crew. He knew, from experience, which of the men were steady swimmers and which needed assistance. He was giving instructions, but the strident moment had passed. In fact, he seemed subdued. I took the pack of Augustin, the botanist, who preferred climbing through the thicket to swimming under it, and Fay came behind all of us as sweeper. He even brought my sleeping pad, which had gotten unpacked during some emergency reshuffling of the loads and been temporarily stowed in a tree. He handed it back to me dry.

By 12:40 p.m. we stood on the west bank, wringing out our shirts (except for Fay, still shirtless), checking our packs for leakage, basking in the sunshine—rare sunshine!—that blessed us there through a canopy gap. Flush with nervous relief, we joked and relaxed. We were pleased with ourselves for having wiggled through what might be, we hoped, the last of the dire obstacles. Emmanuel lifted a sodden 10-pound bag of rice from his pack, letting the pale milky water drain out. Nick labeled rolls of film. Sophiano had a smoke. Fay, head down, quietly wrote in his yellow water-proof notebook.

And then, without comment, without any speech of further remon-strance let alone congratulations, Fay detached himself crisply from our breezy mood. He glanced at his wrist compass. He turned toward the forest and stabbed out his arm, giving the usual signal to Bebe: *That* way. Duti-fully, Bebe stepped out and began cutting trail. Fay walked.

Snatching up my pack, holstering my notebook, I followed. I was startled by his brusqueness, but I wanted to stay at Fay's heels. Maybe, in the aftermath, he would loosen up and commit a personal revelation. Maybe he'd put his outburst in context. Or maybe he'd just encounter something interesting—a Gaboon viper, a gorilla, a dwarf crocodile—that I'd hate to miss. The rest of the party were left behind to think what they might think, to feel what they might feel, to gather themselves at their own pace.

At 1:11 p.m. on Day 453, Fay paused to record the next datum: elephant dung, old. Then again, without speaking, he walked. A hard man, a savvy leader, a flouter of pieties, a solitary soul, a conscientious scientist, a fierce partisan of tropical forest, a keen judge of human limits, he had work to do—not much work remaining now, but some. He couldn't celebrate yet. He was still three days from the beach.

· ·

He made it. Fay reached his finish line on Day 456, which in the terms of the wider world was December 18, 2000. During the final hour of walking through flooded forest, we had heard the roar of surf; we had smelled the salt. Crab holes, as well as wormholes, perforated the mud across which we slogged. The canopy thinned, the trees here were small and scrubby, and we felt the sea breeze. Then we burst out through the last band of brush and there it was: the Atlantic Ocean.

Nearby on the sand were two rusted-out shipping containers, a landmark we knew: They were grounded just two or three miles up the beach from a staging site called Petit Loango South Camp, from which Nick and I had rejoined Fay's march two weeks earlier. The loop was now closed. It seemed amazing that Fay had walked 2,000 miles through African forest, following a compass line, and reached the ocean with that proximity—two or three miles—to his target. I heard him say to himself: "Wow."

Among his first acts was to call his father by satellite phone from the beach, and it's the only time I've heard the timbre of tears in Mike Fay's voice.

Years afterward, I would sometimes wonder about the "data set" upon which he had put such importance—that continuous (except for the Mouko Gap) aggregation of minutely recorded observations across space and time, collected in his many dozens of yellow notebooks and his varied forms of digital storage, and about the website he intended to establish, making available those data, in which people were "going to see very clear patterns," such as the correlation between douka trees left standing to their maturity and a healthy resident elephant population. If that database exists, it's not easy to find. But maybe that doesn't matter.

Maybe the sheer mass of gathered and systematized and searchable data is not the most important or valuable "deliverable" of the Megatransect, notwithstanding Fay's dedication to that dimension of the work. Maybe his fierce dedication itself, and the narrative it drove, and the anecdotes and subjective observations it yielded, were the most potent and consequential yields of this epic expedition, imagined into existence by a single man, and performed by him, under often lonely conditions but with the help and support of many others. There are immediate impacts of the Megatransect for conservation in Gabon (see "A Country to Discover," pages 81–96), and that's what they seem to imply.

Decades afterward, I still encounter people who recall reading about and glimpsing (in those amazing photos by Nick) the Megatransect, and I meet field biologists and conservationists who were inspired by its audacious and hungering scope (see "Saving the Okavango," pages 263–274). What they most remember, I hope, is that it happened in its fullness because one stubborn man treasured the heartbeat of the wild—as it beat within those forests of Central Africa that he knew well, loved well, and wanted to help save, in his own highly data-informed way. It happened because he wanted us to share the privilege of seeing them and hearing their pulse, if not in person, then at least vicariously, gazing over his bony, naked shoulders and listening through his keen ears.

A COUNTRY
TO DISCOVER

·┤·┤·┤·┤·

One Signature, 13 New Parks

I went home from the Megatransect, after those final exhilarating days, carrying some source of new heat in my blood. That's not a fancy metaphor; I mean literally. It didn't manifest until a few days after I was back in Montana, by coincidence on the same day I saw my doctor for an annual checkup. I had begun to feel a little feverish but didn't think much about that. I had a blood sample drawn in the lab, for cholesterol count and other routine screening, and went back upstairs to the doctor, my friend Bob Hathaway. He told me: I hope you have no plans for tonight, because I'm admitting you to the hospital; you're about to get very sick. He had seen the numbers of my elevated white cell count, and he was right. While I answered the routine questions of the admitting clerk, I began to shake with chills. Have you traveled recently? Yes, I was in Africa three days ago. Then I added, making a foolish joke: It's probably just a touch of Ebola.

Planked out on a bed, an antibiotic drip in my arm as prescribed by Hathaway, I did poorly for a few hours. My wife of the time (soon to be my ex-wife, by her choosing, but today still a good friend) remained with me to be supportive through that evening. After my temperature spiked at 105.6°, and because of my stupid Ebola comment, the nurses began wearing masks when they came into my room. But the antibiotic slowly conquered whatever bacteria swam in my blood. The medics never did identify the bug but, late that evening, my fever broke. I walked out next day. You saved me, I told Hathaway. That's just a hypothesis, he said.

Soon it was time to go back to Africa. National Geographic had more stories related to the Megatransect that they wanted Nick Nichols and me to do. Two of them were conservation success stories—it's always a pleasure to be able to report good news—and the third was another manic adventure with Fay, this time involving a small airplane in a dust storm at low elevation above the Sahara, a combination I don't recommend. But first, a return trip to Gabon.

On the morning of August 1, 2002, in Libreville, Gabon, President El Hadj Omar Bongo summoned his ministers to an urgent meeting. Almost no one except Mr. Bongo knew what was up. Did the government face some sudden financial crunch—related, maybe, to falling petroleum revenues and rising deficits? Was there an international crisis, putting all Africa and the rest of the world on nervous alert? Had civil war broken out again somewhere in the region, Central Africa, within which Omar Bongo, during his 35-year incumbency, had earned a certain reputation as a peacemaker? Would the president undertake a mission of mediation? Even as his ministers gathered in the cabinet room of the presidential palace, they had no idea what the day's business would be.

Adding to their puzzlement was the fact that three outsiders had also turned up for the meeting—a British biologist named Lee White, employed by the Wildlife Conservation Society (WCS) of New York as head of its Gabon program; a Cameroonian biologist named Andre Kamdem Toham, based in Libreville for the World Wildlife Fund; and the American ecologist and explorer J. Michael Fay, a WCS employee more familiar to some of those present as the "man who walked across Gabon." The minister of tourism turned to White, an acquaintance, and asked: "What are *you* doing here?"

The cabinet room was an impressive chamber, big as a tennis court, stately as a church, with two great mahogany tables running up the center. At the front was a raised presidential podium, like a postmodern, minimalist throne. Each of the tables was partitioned into ministerial cubicles equipped with telephones and other electronic communications gear. Large plasma video screens faced the tables for audiovisual briefings, with a separate screen positioned to serve the podium. The ministers took their assigned places. After a slight delay, while White and Fay struggled hastily to patch a laptop computer into the room's system, the president entered, a self-possessed man with a wide mustache and a warm smile, looking dapper in a bright yellow business suit. He said nothing. He sat down and, with a nod, signaled his minister of forest economy, Émile Doumba, to start the proceedings. Doumba announced simply that Dr. Fay and Dr. White would address the group on a matter of high interest to the president.

"And so I just launch into my dog and pony show," Mike Fay said, recounting the scene during one of our quiet talks in Libreville months later, his wounds from a recent elephant goring now nearly healed, his zeal undampened by that near-death experience. "The president had a little TV screen in front of his face, and he's staring into it, you know, intently." The ministers soon were engrossed too. Fay, better adapted to bushwhacking through swamps than to cabinet-level politicking, was wearing a jacket and tie borrowed that morning from Lee White's closet. He had brought his laptop, as he carried it everywhere, stuffed into a daypack. The summons to him and White had come on short notice, and the rushed distraction of solving the computer-compatibility problem had left him little time to gather his thoughts about what he would say. But, having delivered variants of the same spiel already to so many and such various audiences, he wasn't shy about winging it before a council of ministers.

Clicking through a series of striking photos (many of them shot by Nick Nichols for the Megatransect stories) in a PowerPoint presentation, Fay described the extraordinary biological riches residing in the trackless forests, the remote mountains, the inland and coastal waters of Gabon, and the extraordinary opportunity—an *economic* opportunity as well as a conservation opportunity, considering the potential earnings from tourism—that might be seized by protecting those riches within a network of national parks. *Click:* forest elephant, stern and alert. *Click:* humpback whale, breaching skyward like a frisky trout. *Click:* Gaboon viper, its big coppery head so close to the lens you could almost feel the flick of its tongue. *Click:* granite inselberg, like a great igneous gumdrop, protruding above forest canopy. *Click:* bulge-eyed hippopotamus, almost unrecognizably strange and serene amid blue water, riding a wave along the Atlantic coast. "Of course, everyone is blown away by the surfing hippo," Fay told me. This is Gabon, he reminded the ministers. This is your country, like none other on Earth.

"And then Lee gets up," Fay said. "He's like the icing on the cake, because he's got video." White's video collage depicted many of the same creatures and places, except that this time the elephants, hippos, and whales were in motion, flickering across the plasma screens that served as windows to Gabonese wonders lying not far beyond the walls of that room. From a clearing amid the vast central forest, a place known as Langoué Bai, undiscovered until Fay walked through it, there was a startlingly intimate sequence of a female gorilla as she suckled, kissed, and dandled her infant. From the faunal reserve called Lopé, just beyond the Chaillu Massif, came the sight of hundreds of mandrills (large monkeys of the species *Mandrillus sphinx,* with striking blue-and-red faces) in full sprint across a savanna.

Some of White's video (notably the maternal gorilla scene, shot by a visiting activist named Sam LaBudde) was so affecting that he let it roll silently, without commentary. By the time he finished, White's presentation plus Fay's had taken more than an hour, and the captive-audience ministers, according to Fay, must have been wondering, *Aren't these guys ever going to shut up?*

Not quite yet. With the president's permission, White launched into a coda, extolling the same grand idea Fay mentioned: a network of national parks.

One by one he described them—13 magnificently wild areas, 13 prospective parks. They ranged from the seacoast at Gabon's southwestern extremity (Mayumba, a potential marine park) to the inselbergs of Minkébé in the country's northeastern corner. They included the fog-topped mountains northeast of Libreville (Monts de Cristal, with their inordinate botanical diversity) and the pristine forest surrounding Langoué Bai in the upper Ivindo River drainage, harboring big-tusked elephants and unwary gorillas, and gated by tall waterfalls on the Ivindo and several tributaries. It was a menu of options representing a wide variety of species and landscapes, rich ecosystems, each area meriting consideration for national park status—someday, perhaps. White could speak expertly about these areas, having directed a comprehensive evaluation of them (in collaboration with Andre Kamdem Toham and aided by various Gabonese partners) during the preceding two years, with the encouragement of one Richard Onouviet, formerly the forestry minister, who now held another portfolio in the cabinet.

Concluding, White showed a map of Gabon. Outlined like giant amoebas were the 13 candidate areas. This is what we think you should do, he told the president and all the president's men. We respectfully recommend that, to conserve biological diversity and promote ecotourism in Gabon, you create such a network of national parks.

TRUTH BE KNOWN, White and his colleagues were reaching for pie in the sky. A multiple-parks network was the long-term goal, and 13 components seemed the best-case result for the future. A more immediate objective, far less ambitious but still difficult enough, was to get park status for just the Lopé Reserve. Lopé was one of several Gabonese areas that, as *réserves de faune,* already enjoyed some protection; but as a national park supported by full management and enforcement mechanisms, with no hunting allowed and no timber extraction, it would be something quite different. Lopé was especially close to White's heart because of his dozen years as a researcher there. With its mandrills, elephants, two decades of gorilla and chimpanzee studies, mysterious archaeological sites suggesting a sizable human population that vanished some centuries ago (the same hypothesis that Fay had mentioned when he showed me the old oil-palm nuts buried in the Mopo River), and first-class hotel, it contained a wealth of attractions for Gabonese vacationers, international tourists, and scientists. If any of the 13 areas stood a chance to become Gabon's first modern protected area, it was Lopé. In fact, White and his two colleagues in the cabinet room hoped that President Bongo might be ready—maybe that very day—to sign a decree establishing Lopé National Park.

No one had interrupted their pitch with so much as a question. Afterward, no one spoke. White, Fay, and Toham sat politely. "It was over," Fay recalled, "and the president was beckoning for something." Minister Doumba went to the podium with an elegant folder. Opening it, he showed the president a document—the Lopé establishment decree. The president browsed it, then shook his head. No, he murmured to Doumba. No, that's not what I want. He didn't address the full group, but obviously something wasn't right.

Confusion, consternation, embarrassment. Going to the aid of Doumba, Fay invited himself onto the presidential podium and glanced over the presidential shoulder. Yes sir, that's it, that's the Lopé decree, he affirmed. Fay could intervene so presumptuously because, based on a single previous meeting and on the media coverage of his Megatransect, President Bongo had taken a shine to him—the crazed American, the wild child who footed his way across all those nearly impassable forests and swamps, who sat half naked atop an inselberg (as captured in one stunning photo by Nick) contemplating the jungle below, who brought back images and tales of a Gabon that Omar Bongo himself hadn't known existed. But even Fay couldn't appease Mr. Bongo at this moment.

Frustrated, the president spoke. *I want the whole thing,* he said. *Not just Lopé. I want the network.* Minister Doumba, poor man, was unprepared for this leap. What network? *I want the network they just described,* said the president. There is no verbatim transcript but by all accounts, he made himself clear. *I want 13 decrees establishing 13 new national parks. I meant to sign them today. Get on it.* And then, with a faint wave to Fay and White, barely a word to anyone else, President Bongo strolled out of the room.

The others remained, looking stunned and vaguely bewildered. "It was kind of an anticlimax," Fay told me later. "We thought, My god, what happened there?"

Elvis had left the building. The music had been transcendent but there were no autographs.

ON AN AFTERNOON in January 2003, near the start of the long rainy season, Lee White and I took off in a small plane from a grass airstrip in central

Gabon and headed eastward, 600 feet above the treetops. Beneath us, in a paisley pattern of savanna clearings, strips of gallery forest along streams, and bosquets (patches of isolated forest), lay the area formerly known as Réserve de Lopé-Okanda. The verdant forest patches and strips were spackled with bright globes of sumac red, auburn, and bloodshot orange, indicating the scattered individuals of a single tree species, *Lophira alata*, collectively marking the seasonal change with their rubicund new vegetation. It was a delicate spectacle, like the Smoky Mountains in early October when the maples have gone crimson but the buckeyes and oaks are still green. Yes, the *Lophira,* remarked White, they do seem uncommonly well synchronized this year. An intense man with an unassuming manner, laconic but not blasé, he sat at an open baggage hatch in the back of the plane, without a seat belt or harness, shooting video of the pretty trees.

We passed above the Offoué River, a modest squiggle of brown, near its confluence with the much larger Ogooué River. The Ogooué, one of Central Africa's great waterways, sharing a divide with the Congo River, oozed seaward from the Gabonese interior like an enormous runnel of gravy. The little Offoué serves as the east boundary, and the mighty Ogooué the north, of what is now Lopé National Park. The decree of establishment for Lopé, along with 12 other such decrees, was signed by President Bongo on August 30, 2002, less than a month after the cabinet-room meeting.

The ink was dry and the parks network was reality—at least on paper, as a matter of law. This 13-park initiative, which Mr. Bongo himself announced a month later at the World Summit on Sustainable Development in Johannesburg, potentially (mark that word) constituted one of the most significant conservation actions since March 1872, when another president, Ulysses S. Grant, signed a bill from the American Congress establishing Yellowstone National Park, the world's first. The new Gabonese parks network reflected a visionary decision grounded in economic pragmatism. After decades of heavy reliance on petroleum and timber industries, Mr. Bongo had told the delegates in Johannesburg, "We are left with little oil in the ground, a fragmented forest, dwindling income, and a burden of debt." The next growth sector of his nation's economy, he vowed, would be "one based on enjoying, not extracting, natural resources."

Whether or not the Gabonese enterprise achieved that potential would depend—as Mike Fay and Lee White well knew—on the rigor and sagacity of the follow-through. And the follow-through would depend partly on the strength of international assistance. In that arena, early signs were hopeful. During the same week as Mr. Bongo's announcement, U.S. Secretary of State Colin Powell also visited the Johannesburg summit and made a complementary declaration: The United States intended to contribute $53 million within a four-year period to a collaborative effort called the Congo Basin Forest Partnership, in support of natural resource management in six countries of the Central African region. A healthy share of the American money would likely go to Gabon. With additional help from other developed nations, including France, Germany, and Japan, and from some nongovernmental organizations (notably the Wildlife Conservation Society, Conservation International, and the World Wildlife Fund), Gabon would be able to create the training programs, the infrastructure developments, the management and enforcement regimes necessary to make the parks gambit a tangible success, both in economic terms (through tourism) and for conservation. If that success did come, it would be huge.

Lopé National Park alone comprised 1,919 square miles of savanna and forest landscape, and the entire network totaled 11,294 square miles, or 11 percent of the land surface of the country. Percentagewise this put Gabon neck and neck with Costa Rica, whose national parks support a thriving tourism industry. But Gabon is five times as large as Costa Rica, and its equatorial forests and wetlands are known to be teeming with life. The sheer sum of natural assets within its 13 new parks, including known species of high concern (such as forest elephants, western lowland gorillas, dwarf crocodiles, chimpanzees) as well as animal and plant forms yet undiscovered, was incalculably large. With the president's decision in August 2002, Gabon was pledged to become one of the world's leading stewards of biological diversity.

LOPÉ WAS A central piece of the parks network, not just geographically but also because of its well-established research station and its classy hotel. But our January reconnaissance flight took White and me eastward to another park, Ivindo, one of the least known and most intriguing of the 13. Still at low elevation, we followed the Ogooué upstream into Gabon's deep interior, the river's riffles and small islands slipping behind us, its banks heavily forested except for the occasional glimpse of a railroad line or a tawny clay road.

Once we passed over the town of Booué, with its airstrip and logging yard, we saw almost no signs of human presence. The Ivindo River, a major Ogooué tributary, was easy to recognize—a big black-water channel pouring in from the north, dark with tannins leached from detrital mulch in the swamps and seasonally flooded forests such as Fay and his crew had traversed. From overhead, as we crossed the mixed zone of the waters, I noticed that the black disappeared quickly into the brown, like some precious decoction of wildness diluted into a world of muddled striving and erosion. You can't bottle that stuff—another reason the world needs national parks. Black water without the swamps and slow seeps that generate it, the creatures that lurk in it, the time and repose necessary to distill it, is just cold tea.

Stay with the Ogooué, I told our pilot through the headphone radio. Twenty miles on we spotted what we were looking for: another black-water river, smaller, this one known as the Djidji. We followed it upstream, the plane carving to and fro, along a gently undulant approximation of the river's serpentine course. We gazed down at one set of minor chutes—a rocky cascade totaling 40 feet of vertical drop—and soon afterward passed the invisible boundary into Ivindo National Park.

The next set of chutes appeared major even from our vantage, hundreds of feet above. Abruptly, from a lip of quiet water screened by trees, the Djidji River dropped nearly 200 feet, its volume split into five fingers that clenched down over the rocky face like a grasping hand, each finger a frothy channel punctuated by ledge holes and rooster tails, plummeting to an explosion of foam at the bottom. We circled. We ogled. After three circuits we continued upstream, where the river's surface again was as sleek as an ebony table. The chutes of the Djidji seemed to mark an escarpment of some sort, above which that river winds sedately across a flat, thickly forested plateau. All we could see beneath us, around us, to the horizon in every direction, was unbroken canopy in its thousand shades of green and, through it, a thin slash of black.

Langoué Bai, foremost of all the hidden clearings Fay investigated during his march, with its concentrations of elephants and gorillas, lay dozens of miles to the south. Lee White himself had led the field team that hiked to that bai, after Fay had put it on the map, and it was White who established a continuous monitoring effort. We had reached it easily today coming upstream beyond the chutes by plane, but the land approach to Langoué, via an old road and then three days of hard bushwhacking, was more arduous, and angled in from another direction. White hadn't explored these mysterious precincts of the upper Djidji River—he had flown over them, yes, but never gotten on the ground—and he now shared my curiosity about what was down there. Early observations at Langoué, he told me, suggested that the elephants drawn to the bai (for succulent vegetation, water, salt, or whatever) made some sort of seasonal migration, moving elsewhere for other ripening resources. Anyway, they disappeared from Langoué Bai when the rains ended. Where did they go? Our guess, he said, is that they come here during dry season to the marshy, provident flatlands of the upper Djidji. This meant it might have been serving as the last unprobed sanctum of Gabon's biggest tuskers.

But no one knew that for fact. The work of data gathering at Langoué Bai had barely begun, and the exploration of the surrounding watersheds, including the Djidji River above the chutes, was another urgently tantalizing task on a list of many. Ivindo National Park, like some of the others, was still a black box of uninventoried treasures.

THE GABON PARKS saga, like the chutes of the Djidji, entailed several split but converging branches. Some of them meandered through channels far from Libreville. Months before Omar Bongo's surprise announcement, a diverse cast of players had been performing varied roles in a complicated narrative, with scenes and subplots set in Washington, D.C., Paris, and elsewhere, that led toward a two-tiered result: Gabon's bold initiative and, as its wider context, the Congo Basin Forest Partnership.

The D.C. subplot included a significant moment in the early days of George W. Bush's presidency when, on June 4, 2001, Walter H. Kansteiner III was sworn in as assistant secretary of state for African affairs. Kansteiner had been a commodity trader with a long-standing professional interest in African products (cocoa, coffee, sugar) and a personal interest in African landscapes and issues, before taking an earlier post under an earlier President Bush. He served in the administration of George Herbert Walker Bush during 1989–1993, on the staff of the National Security Council. At this swearing-in ceremony for the assistant secretaryship, Kansteiner spoke of five priorities he felt should guide American policy toward Africa: improved standards of living, democratic institutions, support for the fight against AIDS, conflict resolution, and "the environment," by which vague term he meant not just breathable air and drinkable water but also the conservation of biological diversity and intact ecosystems. How to deal with such vast concerns? Kansteiner was open to good ideas where he might find them.

Another crucial player was Dave Barron, a veteran behind-the-scenes networker who functioned as a government-affairs adviser to several African leaders, including President Omar Bongo. Through Barron, Kansteiner met Mike Fay and heard about some urgent possibilities that existed in Gabon. Grasping that those possibilities were both real and large, Kansteiner provided support, first in the form of seed money, later by bringing Gabon's initiative to the attention of his boss, Colin Powell. Also through Dave Barron, Fay got the attention of another key State Department honcho, John Turner, the assistant secretary for oceans and international environmental and scientific affairs. Kansteiner, Turner, their deputies and staffs, and the newly appointed American ambassador to Gabon, Ken Moorefield, all became energized toward involving the Bush administration somehow in support of conservation in the Central African region.

Barron worked meanwhile to bring some allies from Congress into the mix. On short notice he organized a private dinner at the Capitol Hill Club for about 60 people, at which the wine was good, the food was decent, and the marquee performers were Mike Fay and Nick Nichols, who together did what Barron recalls (anticipating Fay's own self-deprecating metaphor) as "a brilliant dog and pony show." Like the later one for President Bongo's cabinet, this DPS featured soulful gorillas in place of clever dogs, hippos bodysurfing in the Atlantic breakers near Petit Loango instead of trick ponies. Although it was a business-suit sort of event in a tony Washington club, Fay went tieless and jacketless in a wrinkled plaid shirt. The evening reached an inspirational climax when, after Fay had spoken again about ecotourism economics, Representative Clay Shaw of Florida stood up and gave impassioned testimony to the conviction that, pragmatics aside, America should nurture conservation efforts in Africa because it was flat-out the right thing to do. When the dinner ended, there was a consensus of certitude that something should, could, and would be done.

American commitment to the Congo Basin Forest Partnership took form quietly during the summer of 2002. Mike Fay stayed closely involved, drafting memos on his laptop even during a week spent chimpanzee-watching with Jane Goodall at a remote forest camp in the Republic of the Congo (see "Goualougo and Friends," pages 97–107) and emailing them off to D.C. by satellite phone through a gap in the forest canopy. It all came to culmination when Omar Bongo and Colin Powell made their consecutive announcements in early September 2002 in Johannesburg.

For the George W. Bush administration, pledging $53 million for Congo forest protection may have seemed both an honest, generous act of principle and a relatively cheap way of blunting criticisms—from among the assembled delegates at Johannesburg and around the world—that its postures on environmental agreements and its performance on conservation issues had been almost uniformly truculent and bad. To reaffirm his own earnestness, Secretary Powell stopped in Gabon before flying home, for a cordial meeting in Libreville to congratulate President Bongo and then a brief walk through the woods with Mike Fay.

The coastal forest where Powell and Fay took their stroll lies within another of the new national parks, Pongara. Whatever convergence of politics and principles brought these two unusual men together, in that wild place, on that day, didn't matter so much as the question of what tangible results the Congo Basin Forest Partnership would yield, not just in Gabon but also among the neighboring countries—Cameroon, Equatorial Guinea, the Central African Republic, the Republic of the Congo, and the Democratic Republic of the Congo. Colin Powell's outing with Mike Fay was less than a transect (let alone a megatransect) and perhaps more than a photo op. It seemed to be a signal moment in Washington's awareness that the African continent encompasses an extraordinary richness of biological diversity as well as an extraordinary richness of human cultures, and that wealthy governments of North America and Europe were well advised to invest friendly support in both. It was a hike to the edge of the wild, and this time even the secretary of state wasn't wearing a jacket.

DURING MY VISIT to Gabon that January, four months after the parks declaration, President Bongo was occupied with regional diplomacy, helping mediate a fierce civil conflict in Ivory Coast. The newspapers were full of it. The presidential jet zoomed in and out of Libreville. Unable to reach the busy president, I spoke instead with some of his ministers and other high advisers, among whom the most engagingly candid was the minister of defense, Ali Ben Bongo, who happened to be one of the president's sons. (He would also become his father's successor, ascending to the presidency by election after Omar Bongo's death in 2009.) Educated in Paris, Ali Ben was then a heavyset man in his middle 40s who wore his family status and professional role lightly. I tagged along in his Super Puma helicopter during a daylong inspection visit to another of the new parks—Loango, along the southwestern coast—and shared a casual lunch overlooking the water with him, his wife, Mike Fay, and a small entourage.

Then the Super Puma carried us off again, farther down the coast, with Fay at the minister's elbow pointing out elephants and buffalo amid the forest clearings. In the back of the chopper, the minister's chief of staff tapped me excitedly on the shoulder and pointed to movement in a small clump of trees: *Regardez, Monsieur David, chimpanzees!* Appreciating nature had become a national priority.

At the end of the day, Ali Ben Bongo and I sat aside for a few minutes of private talk. Not long before the cabinet-room meeting back in August, Ali Ben told me, his father gave him an inkling of what to expect. The president had seen all those new photos of Gabonese wildlife, he had met Fay, he had watched the National Geographic Television documentary *Africa Extreme* about the Megatransect. The images came as revelations. Because Gabon was still 75 percent forested, with a small human population (only 1.3 million in 2003) and few roads, the president's family, like most of the affluent class, had done their traveling across it mainly by airplane. They seldom drove, Ali Ben said, and they certainly didn't walk through the forests and swamps. But then his father saw what Fay had seen—and he decided to do something.

The next crucial steps would be to organize effective management structures, train people for those management roles, establish real protection for the areas, and help Gabon's populace understand the importance of this initiative. Financial support, coming from the United States and other friends through the Congo Basin Forest Partnership, would be vital. But the partners had to realize, Ali Ben said, that Gabon itself—no one else—would define the goals and the methods of the parks initiative.

The defense minister added a personal note. When he was a teenager, raised in the palace and privileged to travel, he once visited the San Diego Zoo. There, for the first time in his life, he saw a Gaboon viper. It fascinated him and piqued his pride, but it also triggered an unease about the disconnection between creature and place. To see this formidable Gabonese snake, he'd had to go to California? (That seemed poignant and incongruous to me too, because I'd had the pleasure of encountering Gaboon vipers in the wild, during my walks with Fay, each one curled and lurking and huge, so freaking close to our line of march that we almost stepped on it.) "We don't want to get to a situation," the minister and presidential son (and future president) said now, "where we'd have to go to Europe, or the U.S., to see in zoos some of our own wildlife that have become extinct here." Better to preserve what Gabon had been given, in a network of parks such as his father had decreed, and let the world come to visit the snake.

"Imagine," said Ali Ben Bongo, "the third millennium. And we Gabonese still have a country to discover."

....

In June 2009, President Omar Bongo died, in Spain, after treatment for bowel cancer. Ali Ben Bongo (more formally, Ali Bongo Ondimba) was elected to succeed him and sworn into the presidency in October. He was reelected in 2016, amid some accusations of election irregularities and human rights violations. Certainly, it is unusual—and suspect, to observers from other democracies—for a father and son of one family to have held the presidency of a nation continuously for more than 50 years, as père et fils *Bongo have now done. But in at least one matter—the matter of conserving Gabon's great forests, wetlands, and the biological diversity they contain, as well as the marine ecosystem offshore from Gabon—the Bongo presidencies have been very beneficial.*

Seven years after the parks-network declaration, and in the first year of the Ali Bongo Ondimba presidency, Lee White was appointed director of Gabon's National Parks Agency. He served in that position for 10 years, then in 2019 became the country's Minister of Water, Forests, the Sea, and the Environment. It's hard to know exactly how influential White has been in that pair of roles, but underestimation would be a mistake. His appointments certainly reflect both the degree of trust that Bongo Ondimba placed in him and the president's enduring conviction that conservation of Gabon's biological diversity should be a major governing priority—as then Minister of Defense Bongo suggested to me, back in 2002, that it would be.

White has also helped assure that good science be part of good governance. When I asked him recently for news, he sent me a packet of papers, from journals such as Nature, Science, *and* Biological Conservation, *on which he, his scientist-wife Kate Abernethy, his old colleague Mike Fay, and many Gabonese and international scientists are co-authors. One of them described "how we repeated the forest parks process in the ocean." Under the second President Bongo, the govern-ment established a framework called Gabon Bleu, engaging multiple national agencies and attracting support from international funders to create, by 2017, a vast complex of marine parks and aquatic reserves. Some of these extend as far as 200 nautical miles off the country's southwestern coast. Commercial fishing is still permitted in certain designated areas (a relatively small share of the total); other areas are open for subsistence fishing only or protected completely. Sea turtles and whales benefit. Populations of tuna, shrimp, and many other compo-nents of the marine community—commercially targeted or not—are better able to sustain themselves and thrive.*

Another of the attached papers, White told me, showed "how we are winning the war with ivory poachers." It reported on a systematic nationwide survey of Gabon's forest elephant population, which per-suasively suggests that it has increased by roughly 50 percent since the bad old days of the early 1990s. Earlier estimates, back then, put the Gabon elephant population between about 41,600 and 82,000 ani-mals. The new study—which is the first of its kind, in scope and method, for any free-ranging large mammal in Africa—suggests that roughly 95,000 elephants are roaming wild across 96,800 square miles of Gabonese habitat. "The snapshot of forest elephant density distri-bution and abundance we report is good news for the species in Gabon," the paper's authors declare. Poaching remains a concern, and so does the potential for human-elephant conflict in less premeditated forms, as Gabon's human population continues to grow—and as more people live and work within logging concessions, where elephants also seek food and security.

How do you reconcile those growing numbers and needs—human and elephant—when the forest area itself never gets any larger? With care, wisdom, professionalism, and anguish. But if any country can achieve that, Gabon might be it.

GOUALOUGO
AND FRIENDS

A Walk in the Deep Forest
With Jane Goodall

....

Jane Goodall has had a long relationship with the National Geo-
graphic Society, and for a time held the position of Explorer in
Residence, which entailed making herself occasionally available for
worthy Society activities. Mike Fay became an Explorer in Residence
too. In 2002, Fay suggested that Dr. Goodall could be helpful to an
important conservation opportunity in the Republic of the Congo,
near the site where his Megatransect began, if she might be willing
to come, and bring the spotlight of her renown, to the Goualougo
Triangle. The Goualougo (as you've read earlier in this book) is a
wild area of superb chimpanzee habitat adjacent to, but not yet at
that time part of, Nouabalé-Ndoki National Park. So the world's
most famous primatologist and chimpanzee advocate made a foot
pilgrimage with Fay to meet the chimps of the Goualougo, a place
so remote that those animals hadn't yet learned to fear humans. The
Goualougo was so remote, in fact, that its chimps had never heard
of Jane Goodall. Nick Nichols and I went along.

....

I t was an amazing display of bravado from animals of a species not
generally perceived as being fierce. The chimpanzees moved in through
the treetops, hooting and shrieking like a pack of hungry predators on
the hunt. At that moment, in fact, they *were* hungry predators on the hunt.
Diving from limb to limb, gabbling excitedly, they set up a menacing

ruckus. Vines shook. Branches fell. Using their weight and strength, they stirred the canopy like storm winds. Their war whoops were spookier than martial bagpipes on a Scottish moor. Intermittently they paused to crane and ogle, scanning the ground ahead for a glimpse of their prey. Chimps, after all, are not vegetarians; they eat fruit and leaves routinely but relish meat when they can get it. This group had been drawn by the bleating moans of what they took for a duiker (a small forest antelope) in distress— and a distressed duiker, to a group of chimps, represents a potential bounty of protein. Maybe they expected to find a wounded adult, or a newborn fawn, or at least a pile of succulent afterbirth. Anyway, they hadn't yet realized that the duiker bleat was a decoy call, made by a Bambenzele tracker named Youngai, who hunkered quietly amid the understory, waiting for them to come. And they didn't know that beside Youngai sat three other human visitors, each of us thrilled with the privilege of encountering chimpanzees so bold as to mistake us for prey.

As the chimps approached closer, catching sight of us on the ground, their excitement didn't lessen—but it changed. Suddenly they looked surprised and perplexed. We could see ourselves register weirdly on their awareness. They showed no fear, and their hunters' menace had dissolved. Now they were curious. They settled onto limbs just above our heads and lingered there, gawking, chattering, like a gaggle of fascinated schoolchildren getting their first glimpse into a monkey cage. One female chimp held an infant whose large ears stuck far out from its head, glowing amber like a pair of huge, dried apricots whenever they caught backlighting from a shaft of sunlight. I gaped at the little fellow, just a dozen yards above. His face was tranquil, his eyes widened by innocent wonder. He and his mother gaped calmly back.

Dave Morgan, the younger of my two American companions, positioned his spotting scope and began fixing on one chimp after another, looking for facial markings of distinctive identity. The other scientist of our little group was J. Michael Fay. He put his video camera into action.

They were both seizing precious minutes of close contact to document one of the most arresting phenomena to be found in an African forest:

chimpanzees so remotely isolated that they showed no sign of ever having been hunted, or frightened, or otherwise contacted by humans.

Hours later, after we had stumbled through gathering darkness into a swampy campsite, we discovered that the chimps had followed. That night they bedded in treetops just a short stroll away. In the morning they were with us again. We moved slowly through the forest, and a day later, one chimp approached by foot to within 20 paces of our morning camp-fire. He stood behind a tree, peering nosily. Maybe he fancied the smell of coffee.

The date was September 28, 1999. It happened also to be Day 9 of the Megatransect, Mike Fay's epic survey hike across Central Africa. (This is the same chimp encounter I describe in "The Long Follow," pages 19–40.) Our location was deep in the northeastern corner of the Republic of the Congo, within a spectacularly pristine wedge of forest known as the Goualougo Triangle. From this point Fay would keep walking—and walking and walking—until he reached the Atlantic Ocean, 447 days later. Dave Morgan would remain behind, continuing his study of the Goualougo chimps. None of us foresaw that three years later we'd be together again, joined in our search for another glimpse of these trusting animals by the world's foremost chimpanzee maven, Jane Goodall.

THE RETURN TRIP occurred in 2002, just weeks before the World Summit on Sustainable Development convened in Johannesburg. Goodall was committed to attend the big gathering, which would include presidents, cabinet ministers, scientists, conservationists, development experts, and activists from roughly 190 countries. But in the meantime, she had made space in her schedule for a quiet walk with a few kindred souls in the Congo forest. It seemed a good time and a good place to contemplate the future prospects—if any—for the survival of viable chimpanzee populations within large intact blocks of African forest.

Primate taxonomists currently recognize four subspecies of the common chimpanzee *(Pan troglodytes)*, spanning a distributional range from Senegal on the west coast of Africa, through Gabon and the Republic of the Congo, into Uganda and southwestern Tanzania in the east. That is, they stretch straight across middle Africa, but not across the great Congo River. (As to their relatives on the far side of that barrier, see "The Left Bank Ape," pages 173–185.) At one time, that range may have been nearly continuous, but nowadays the forest areas still occupied by chimpanzees present a map pattern of discontinuous remnants, small patches, and dots. Under pressure from humans, the species has suffered population decline, habitat fragmentation, and in some places local extinction. Although there were probably more than a million chimps in Africa a century ago, no more than about 200,000 (and possibly far fewer) survive today.

In many areas where humans and chimpanzees have come into contact, hungry people treated them as just another form of bush meat, and chimps learned that the naked ape, us, can be lethally dangerous. When the source of conflict wasn't meat hunting or the capture of infant chimps for pet trade or for zoos, it was habitat destruction. Humans felled trees and cleared land for settlements and agriculture, wrecking the chimpanzee world, driving chimps away, leaving them marooned within remnants of habitat—little patches of forest such as the Gombe Stream Game Reserve in Tanzania, where Jane Goodall began her research career back in the summer of 1960.

Although Gombe is now a national park, it's a tiny one, barely more than 13 square miles in area, bordered by Lake Tanganyika on one side and by deforestation along nearly all the rest of its perimeter. Its resident chimpanzees have been studied continuously since Goodall arrived. To her, each chimp has always been an individual, worthy of individual attention and concern—that is, in some sense a *person*—and many of those individuals became well known through her writings. Readers worldwide remember her portraits of ragged-eared Flo, trusting David Greybeard, murderous Passion, and others. Their individual fame tended to obscure the reality that, collectively, Gombe's chimps are few in total and perilously isolated.

The park holds about a hundred chimpanzees, which (as the modern science of conservation biology warns us) may not be a viable population. That is, it may be too small to renew itself indefinitely. Inbreeding could cause trouble. An epidemic might wipe out half the number, after which a drought, fire, or some other natural catastrophe might reduce the other half to a still lower level from which recovery is unlikely. Even without further human incursion, even without poaching or persecution, a population so small and isolated faces considerable jeopardy of extinction. Jane Goodall recognized that dire prospect and was taking steps to try to avert it.

The chimps of the Goualougo Triangle, about a thousand miles northwest of Gombe, inhabit a much different set of circumstances and possibilities. Their peculiarity was first noticed by Fay himself in 1990, when he and a Congolese colleague, Marcellin Agnagna, made a series of exploratory hikes to survey forest elephant populations for a study sponsored by Fay's employer, the Wildlife Conservation Society (WCS). What seemed peculiar was this: These chimps didn't flee from the sound, smell, or sight of people. On the contrary, they sometimes approached, gawking, confident, and apparently fascinated. They were naive about any potential danger from humans—which suggested that they had never before experienced contact with people. The Goualougo was at that time so remote (unreachable by road, bush plane, or human trail) and so unsullied (there were not even any machete cuts of the sort left by Aka forest people of the adjacent region) that one could plausibly imagine it had gone unvisited by people for . . . well, maybe for centuries.

The exploratory treks by Fay and Agnagna led to the establishment, in 1993, of Nouabalé-Ndoki National Park, of which Fay became the first director. But when the park boundaries were drawn, in a process involving political compromise, the Goualougo Triangle wasn't included. The cone-shaped piece of spectacular chimpanzee habitat and old-growth hardwoods, delineated by the convergence of the Ndoki and the Goualougo Rivers, was left dangling beneath the park's southern boundary like a precious but vulnerable appendage. Instead of receiving park protection, it remained held within a timber concession. Eventually, in that status, it would likely be logged. In the meantime, remembering those eerily brazen chimps, Fay tried to learn about them—with as little disruption as possible—before it was too late. In time, the assignment fell to Dave Morgan.

Morgan had studied biology at Western Carolina University and then worked several years as a zookeeper at Busch Gardens in Tampa, Florida. His job there was to feed and tend the captive gorillas and chimps. He had never been to Africa, let alone seen an ape in the wild, until Mike Fay recruited him. In late 1995, Morgan came to Nouabalé-Ndoki as a volunteer assistant on a WCS gorilla-monitoring project within the park. After having proved himself hardy and capable, he wrote a proposal for a pilot study of the chimpanzees of the Goualougo. Fay, increasingly concerned that the Goualougo might soon be logged, arranged funding for the study and set Morgan to it.

On February 24, 1999, Dave Morgan began work, basing himself at a simple field camp in the Goualougo. The chimps as he found them still seemed blithely innocent of the possibility that humans might represent any threat. They sometimes approached to within a few yards, lingering in trees just above, watching him as curiously as he watched them. He counted heads, observed behavior, sketched their faces in his notebook, and when possible captured them on video. By the end of September, despite a month lost for medical reasons (he had been attacked and bitten deeply through the shoulder by a distraught gorilla, an occupational hazard), Morgan had portraits of 93 individual chimps. "The naive behavior" of the Goualougo chimps, he wrote in his pilot-study report, "facilitates the rapid collection of a substantial body of data." And so the study continued.

At the end of 1999, during a visit back in the U.S., Morgan met a young graduate student named Crickette Sanz, then just beginning her work toward a doctorate in physical anthropology. Later she visited the Goualougo, in search of a dissertation project, and liked what she saw. Morgan and Sanz, assisted by a small crew of expert Bangombe woodsmen, pursued the Goualougo chimpanzee study as a joint effort. Their work, coupled with the international attention thrown on the area by Fay's Megatransect expedition, spurred ongoing negotiations between WCS and the logging company, Congolaise Industrielle des Bois, that held the concession. The company has since voluntarily relinquished its timber rights in the Goualougo—a generous as well as a savvy act—so that the area could be annexed to Nouabalé-Ndoki National Park.

As a small complement to other conservation measures taken in this part of Africa, saving the Goualougo Triangle carries major significance. As added to Nouabalé-Ndoki, it enhances the possibility of preserving a very large expanse of continuous chimpanzee habitat and, within that, a sizable interbreeding population of chimpanzees (thousands, rather than merely hundreds or dozens) into the next century and beyond. Securing such a single big area is crucial to the survival of the species, given that so many of Africa's other chimpanzee refuges are, like Gombe, far too small and too isolated to support viable populations.

The chimps of the Goualougo Triangle still enjoy the possibility of an unbounded and genetically robust future. That fact, in addition to their naive attitude toward humans, is what has made them such a focus of interest and concern. But can they remain so naive? If not, then what forms of chastening experience await them? Will they lose their ingenuous curiosity about humans by way of the intrusive attentions of ecotourism, rather than by the lethal traumas of hunting, habitat destruction, and beleaguered insularity? Such questions reflect the real distance—it's more than just land miles—between Gombe and Goualougo.

Naïveté is a delicate, perishable state of being, and in fact the Goualougo chimps had already begun to lose theirs, even within a few years of the Megatransect. Although they hadn't acquired any noticeable fear of humans, their curiosity already seemed less strong and impetuous in 2002 than it had been in 1999. The episodes of excited mutual ogling were less frequent. The limelight of continuous study, even by two such deferential scientists as Morgan and Sanz, seemed to have jaded them slightly. The physicist Werner Heisenberg warned us about this: You can't observe anything closely without affecting it somehow.

ON THE EVENING of the arrival of Jane Goodall, footsore and weary after a long day's slog, and accompanied by Fay and a handful of others (including photographer Nick Nichols and me), Morgan and Sanz were there at the Goualougo field camp to greet her. Night had fallen before the hiking was done, and we found our way down the last thigh-deep channel by headlamp. Stumbling up onto solid ground, we pitched our tents, washed, and reconvened at the campfire for beans and rice.

It had been 10 years since she had walked so far, Jane said. Her blistered soles reflected that fact. Still, at age 68, her signature ponytail now going gray, she had a reservoir of strength to spare—spiritual strength, if not muscular. She seemed invigorated by the sheer joy of being back in a forest full of chimpanzees.

Next morning Jane ventured out onto the Goualougo trails, hoping for a view of the animals Morgan and Sanz had been studying. But it wasn't like the solitary, early days at Gombe. Here, now, she moved at the center of a crowd: a tracker, Morgan, Sanz, Fay, Nick with his unobtrusive little Leica—and that was just the half of it. With each step Jane took, a crew from National Geographic Television shadowed her, hungry to record every word and glance. The forest itself became a TV stage. But she was patient and professional, hitting her mark in every scene, repeating this or that comment when another take was called for, using the television attention as she used all such burdens and opportunities of fame—to get her message out. That message, grossly compressed and presumptuously summarized by me, was: Every individual counts, both among nonhuman animals and among humans, so if you renounce callous anthropocentrism and cruelty, your personal actions will make Earth a better place.

After five days in the forest, it began to seem questionable whether Jane herself, the guest of honor, would have any significant encounter with any chimpanzees whatsoever. One problem was her damaged feet. Although the blisters didn't stop her from walking, they did inconvenience her. Having observed a certain coping measure of mine, she expressed interest, and one morning she let me duct-tape her feet. I covered her blisters and hot spots with little bits of clean paper, and then the paper with my preferred brand of pliable green duct tape, all to fit beneath her sandals—the same method I'd used throughout the Megatransect. It helped. She borrowed my tape for future mornings, did her own taping, and carried on gamely.

Another problem was the sheer collective bustle of such a large group. You don't parade through the woods in a party of 10 if you want to see animals, not even if the animals in question are naive, or habituated, or flat-out deaf. You've got a sound person dangling a boom mike. You've got a camera operator walking backward through the brush. You've got a director who keeps saying, *David, please step out of the shot.* I stepped out, and eventually abandoned the circus altogether, spending a few days exploring other trails with Dave Morgan.

Finally, after most of a week, Jane did get a chance to enjoy what she had come for—three hours in the presence of a relaxed group of chimps as they fed, rested, and otherwise occupied themselves in a *Synsepalum* tree. It wasn't a dramatic encounter. The chimps went about their business, showing no excited curiosity or reciprocal fascination. But it was satisfying to Jane, who saw not just a gaggle of primates but individual creatures, particularized under the names by which Morgan and Sanz had come to know them: the female Maya, her infant daughter, Malia, the female O'Keefe, and a half dozen more.

Later that afternoon, Jane and I sat in the forest discussing the problems facing Gombe, her own years of experience there, and the prospects of an alternate future for the Goualougo. At one point I asked about the difference between concern for individual animals and concern for endangered populations. To her, it's a sterile distinction. "When I'm thinking about some forest being logged, and the bush meat trade," she said, "it isn't just a population of chimps that's going. It's individuals." Destroy individuals of such a species, and you eradicate also "all their wisdom, all their cultures that have been passed down from one generation to the next." After a moment, she added, "I can't separate the loss of a population from the harm of individuals." At Gombe she had known four generations intimately. To the chimps of the Goualougo, she was a stranger. "It does take me back to my childhood dreams," Jane said. "You know, I'm really happy that I got here—in spite of the blisters!" Next morning, on nearly healed feet, she started walking back toward the world.

It worked. Or anyway, it helped. Roughly one year after Jane's visit, and not many months after this story was published, a proclamation from the Republic of the Congo government announced that the Goualougo Triangle, encompassing 25,000 hectares (more than 96 square miles) of remote primary forest, would be added to Nouabalé-Ndoki National Park. That was a start—the proclamation—but the gears of governance and law still had to turn. They did, and the official decree modifying the park boundaries to include the Goualougo came from the president of ROC, Denis Sassou Nguesso, in 2012.

Throughout all this time, David Morgan and Crickette Sanz continued their efforts, along with a succession of colleagues, both Congolese and international, at understanding and protecting the Goualougo chimps. They also extended their research to the western lowland gorilla, which other researchers had studied within the original park boundaries—especially at Mbeli Bai, a large clearing and a favored feeding spot for gorillas, elephants, and other herbivores, where Morgan began his own research career in 1996—but not previously in the Goualougo.

Morgan and Sanz co-direct the Goualougo Triangle Ape Project (GTAP), within which they, their field partners, and their scientific collaborators study topics ranging from chimp tool use to population genetics (of both apes), to sharing of resources and long-term social networks between chimpanzee and gorilla groups whose territories overlap, to the different strategies by which chimps and gorillas meet their feeding and nesting needs in areas where selective timber harvesting occurs (just beyond the Goualougo boundary). The Goualougo thrives today as one of the most precious and diverse enclaves of Congolese forest and one of the world's premier research sites for studying great apes in circumstances very little disturbed by human activity. In fact, it's the only site in Africa, Morgan recently told me, where both chimpanzees and gorillas are followed by human observers daily for collection of standardized data.

In addition, Morgan and Sanz were asked in 2014 to take over management of research and conservation activities at a second site, Mondika, just outside the boundary of Nouabalé-Ndoki National Park, within a rich forest area known as the Djéké Triangle, adjacent to the park's western edge. Mondika has been a research site since the late 1980s, when primatologists from Kyoto University established it. In 1995, Professor Diane Doran of Stony Brook University began a long-term gorilla study there. After her departure and retirement, the Wildlife Conservation Society took over direction of Mondika in 2005, and nine years later Sanz and Morgan assumed responsibility for the work there. They are using the same model of applied research as they have for two decades at Goualougo, Morgan told me, to make the case for a similar outcome: that the Mondika site, with its relatively undisturbed gorilla and chimp populations, be annexed officially to the park. That should happen within a year or two, he hopes.

The opportunity to develop these long-term studies, Sanz and Morgan testify, makes the two sites—Goualougo and Mondika—unique in Central Africa, and contributes knowledge toward addressing conservation threats against apes across equatorial Africa generally. It didn't all begin with Jane Goodall's visit in 2002, but that visit, and Jane's firsthand experience of the Goualougo forest and its chimps, which informed her vocal and faithful endorsement of the work in following years, have been vastly important. Sanz and Morgan aren't alone in considering Jane's visit one of the landmark events in the three-decade history of Nouabalé-Ndoki National Park.

THE
MEGAFLYOVER

·❦·❦·❦·

Mike Fay Goes Airborne
Across Africa

After his time on the Megatransect, and a period of trying to sleep
within four walls in Washington, D.C., while he worked to organize
his data and journals, Mike Fay grew restless. He missed the green
world. So he pitched a tent in Rock Creek Park, just northwest of
downtown D.C., in some secluded thicket unnoticeable to joggers
and muggers, and lived out of that urban encampment for some
months, commuting by foot to his office space at the National Geographic Society. He continued to play a role in conservation efforts
back in Gabon. And then he conceived a new project, a new way of
targeting imperatives for the protection of wild places, more ambitious
even than walking 2,000 miles through Central African forests with
a GPS and a yellow notebook. He would fly.

More precisely, he would megafly. He would aviate like a gadfly
over the whole African continent, at low elevation in a small plane,
tracing a tight gridwork pattern while recording video and other
data, to document the conditions of landscape and the scope of human
impacts on the ground. He called this enterprise the Megaflyover.

Because I'd been involved with the Megatransect experience,
National Geographic's editors asked me to join Mike again, somewhere amid his megaflight, and get aboard that plane. I caught up
with him in the desert of Niger.

Just north of the old caravan town of Agadez, in central Niger, stretches the Aïr Massif, a vast range of cinder gray highlands standing up from the Sahara like a coal barge afloat on an ocean of café au lait. The peaks and plateaus of the Aïr have been shaped over time from a complicated mixture of rock types—including magmatic ring dikes, granitic intrusions, Paleozoic sandstone, and recent flows of lava—but the overall impression they convey can be captured without geologic jargon: big mountains, arid and dark and steep. Their gulches (koris, in the local terminology) are water-carved but brim only with sand in dry season. Old hoof trails, scratched across high ledges, suggest that once this was good habitat for Barbary sheep (Ammotragus lervia), a hardy species now reduced to rarity or extirpated entirely across most of its former North African range. Maybe the habitat is still good, but the sheep seem to have been mostly hunted out. There are no paved roads and few settlements amid these mountains. Apart from four-by-four tracks up the larger koris, the main signs of human presence are igloo-like rock piles sparsely polka-dotting the foothills. Each pile is an ancient grave. The graves are remote, inconspicuous, mostly unopened by pillagers, and best seen from a low-flying plane. That's how Mike Fay saw them, on a mild December morning in 2004, as an heirloom Cessna 182 carrying him and three others approached the northeastern edge of the massif.

The Cessna, showing call letters G-OWCS, was painted scarlet and specially equipped for collecting data. The call letters honored Fay's employer, the Wildlife Conservation Society (WCS), which had supported his varied African labors for 20 years. The plane looked like a toy, or an enameled piñata, but it bore serious purposes, not candy. With a young Austrian pilot named Mario Scherer at the controls, and Fay in the right seat amid a rat's nest of custom-rigged digital hardware and cables, it caressed the topography, circling here, dipping a wing there, rising nervily through high notches to put peaks close at eye level on each side. Mounted in its right door was a high-resolution digital camera that, automatically every 20 seconds, took a vertical shot of the ground. The photos, each tagged with GPS data registering exact time, latitude, longitude, and altitude, were uploaded into a Hewlett-Packard tablet computer on Fay's lap, through which he could add notes. A similar tablet, scrolling out a map along the plane's flight line, rested under his left elbow. There was no in-flight movie. Fay's attention flicked constantly, tirelessly, between the computer screens and the terrain passing below. He wore headphones and a scruffy gray beard.

The plane's interior was as spacious as a Volkswagen Bug. Behind the two seats lurked an 80-liter auxiliary gas tank, like a riser of welded aluminum, useful for long flights over jungle or desert and as a bum rest for anyone rash enough to crawl aboard as a passenger. Jammed beside me on this fool's bench sat a man named Maurice Ascani, small but excitable, claiming little space for his buttocks but much for his personality.

Ascani was a French-born Nigerien with a blunt manner, a deep commitment to Niger's wildlife, and a strong resemblance to the actor Roy Scheider, though perhaps less capable of taking direction. He served as communications officer of a local conservation group called SOS Faune du Niger. Having traveled the country's Saharan outback for more than 30 years, observing and photographing the fauna, making friends among the desert tribes, watching some of the most magnificent species (such as Barbary sheep and that big spiral-horned antelope, the addax) suffer decline, Ascani was well qualified for his role this week as Fay's expert local guide and collaborator. Experience hadn't jaded him. He craned at his window to ogle the mountains as though glimpsing them for the first time. Occasionally he elbowed me to appreciate something—"*Voyez!*"—on his side, or he lurched forward to holler advice at the pilot. One difference between him and us three, besides his disinclination to sit still, was that Ascani knew what to expect as the Cessna neared a certain point known as Arakao.

That point lies at latitude 18° 55' N, longitude 9° 34' E. My map showed it merely as a small hovering label along the east face of the mountains, but Ascani had seen the place firsthand. He had been on the ground at Arakao. He had shot some dramatic images. If he shifted with anticipation as we drew close, I failed to notice, shifting my own sore rump on the aluminum tank.

Then suddenly we were there. Looking from the mountains toward the open desert, we beheld an amazing spectacle: dunes, towering dunes, piled up along the massif's eastern face, like a herd of khaki dinosaurs stopped by a giant stone wall.

SET IN STARK opposition to the dark peaks of Aïr, these dunes are mountains too, of a much different sort: granular, graceful, silkily textured, shaded gently in tones of tan and pale salmon, erected and sculpted into pyramid peaks and razor-edge ridges, swaybacks and rippling slopes, by the winds that have blown them in, grain by grain, across 150 flat miles from north-eastern Niger. Arakao, it turns out, is nothing more than a name for the spot to which those winds deliver their cargo, almost as though they are whistling down a tunnel. Hitting the mountains at a very particular point—the partial cone of an ancient volcano—they swirl, scatter, and lose hold. The sands fall. The dunes rise, some up to 900 feet. And here they linger, bunched and tall, majestic and delicate and dynamic, continually sliding away and continually rebuilt.

With a stylus, Fay tapped a laconic note into his lap tablet: "mountains of luscious sand." His overhead photos would say that and more.

Mario dropped one wing of the Cessna and we circled out over the desert. Gaping back, we saw the dunes with their shadows and edges in bright silhouette before the mountains. Light against dark, smooth against jagged, from this angle they seemed to be pouring down slowly from the highlands like a glacier of sand. We circled again. Three separate GPS units—two for Fay's system, one for the plane—traced our loop. The door camera went *click click click*. With a second camera at the opened window, Fay took handheld shots. Desert wind filled the cockpit. And I began, when Ascani wasn't jabbing my ribs, to shape an inchoate thought: mountains of sand, mountains of data. Metaphor is unscientific, I knew, but then again, I'm not a scientist. Here we had nothing but tiny particles, assembled by a persistent force, yet the collective effect was momentous and grand. As for Fay? He was trying to create his own Arakao, a great dune of information.

IT WAS OUR 10th day of survey flying in Niger, and the 187th day since Fay and his chief pilot, Peter Ragg, departed from an airfield in South Africa on this latest breakneck adventure in ecological reconnaissance, loosely labeled the Megaflyover, for its parallels with Fay's Megatransect. No, the jungle boy hadn't gone soft. Traveling by bush plane rather than on foot, covering hundreds of miles daily rather than half a dozen, and sitting dry in the sky rather than slogging through black-water swamps and thorny thickets didn't represent a change toward safety and comfort. It merely added scope.

Whereas the Megatransect was a single long hike across some of the wildest remaining forests of Central Africa, the Megaflyover was a zigzaggy marathon of low-altitude flights tracing cloverleaf patterns over much of the continent, from Cape Town to Tangier. Despite the differences in mode of travel and geographic reach, the Megaflyover had a similar purpose: to gather abundant, incremental, and systematized data on the state of wild landscapes and the trends of human-caused transformation. Fay's motive wasn't idle curiosity. His aerial enterprise was closely linked with—to some extent inspired by—a major initiative of the Wildlife Conservation Society, known as the Human Footprint project. That project, involving an ambitious program of multidimensional mapping to show gradients of wildness and human impacts around the world, was conceived to help WCS target conservation efforts and funds. Fay himself, a restless individualist with a surprisingly good nose for politics, wanted nothing less than to change the way the world perceives and uses ecosystems and natural resources—starting with perceptions in Washington, D.C. The goal of his Africa Megaflyover, he said, was to convince "the powers that be, in particular the U.S. Congress," that integrating natural resource management into American foreign policy is "a very, very smart thing to do. And a good investment."

Wherever humans live at high population densities, making unsustainable demands on natural systems, he noted, you eventually see ecological breakdown, unmet needs, and tensions that lead toward conflict. Look at Darfur. Look at Rwanda. Look at Zimbabwe. Get beyond the headlines, beyond the tribal and racial animosities, to the resource disputes that underlie them. He was a collector of small facts who liked to think big, and his current line of thinking involved the strategic security issues inextricably linked with water, soil, mineral deposits, flora, fauna, and ecological health. To that end, he conceived the Megaflyover. As a pilot himself, he recognized the value of low-altitude flying to illuminate the realities of land use. A bush plane shows you patterns you'll never perceive from the ground, while allowing flexibly focused coverage ("Let's circle that spot again") and the capture of fine details you can't get from satellite imagery. A modified Cessna 182 was the logical tool. Africa, the continent he knew and loved best after 25 years of working there, was the logical place.

Of course, Africa isn't one place; it's a million places. Its history is as deep as Precambrian bedrock, its landscapes more diverse than those of any other continent on the planet. Nowadays it encompasses 54 countries, hundreds of tribal and ethnic entities, a total population of 1.2 billion humans. It can also be parsed into 104 terrestrial eco-regions (according to another mapping project, this one done by the World Wildlife Fund), each unique in its physical and climatic features and harboring distinct plant and animal communities. Those eco-regions in many cases transcend national boundaries. They range from the Succulent Karoo, in western South Africa and Namibia, to the Saharan Halophytics in northern Algeria. They also include the Western Congolian Swamp Forests, the Itigi-Sumbu Thicket, the Angolan Miombo Woodlands, and many exotic-sounding others. Within or near all these eco-regions live people, at greater and lesser concentrations, whose most elemental struggles and aspirations transcend ecological boundaries as well as national ones, thrumming steadily like the bass notes of a symphony.

Anyone who listens can detect those notes. Africans want better and fuller employment. They want food security and education for their children. They want good governance, free of oppression and corruption. They want fair, sensible arrangements for the management of wild landscapes and natural resources—arrangements chosen and controlled by Africans. They want peace. They're proud to be African as well as proud to be Dogon or Fang or Bambenzele or Tuareg or Samburu or Tutsi, to be Kenyan or

Ghanaian or Gabonese. Directly or indirectly, they suffer from the widespread ravages of AIDS, the pressure of population growth, and the broadly ramifying crush of poverty. Old-fashioned colonialism is mostly gone, but its thefts and damages haven't been well rectified. Increasing urbanization brings rural people toward new enticements, new opportunities, but also toward new disappointments and miseries. In worst-case situations (such as the Second Sudanese Civil War, or the recurrent warfare for control of eastern DRC), political and ethnic conflicts combine with severe natural circumstances to produce masses of refugees, famine, state-condoned persecution, a culture of bloody lawlessness, and even genocide.

Along with the human struggles come human impacts. Although some areas of landscape are less heavily inhabited than they might be, others are overburdened, eroded, blighted by the presence and demands of too many people. Because the African landmass is so large, climate change may affect its interior regions disproportionally, bringing considerably higher temperatures, worse droughts and floods, increased desertification, and new patterns of disease. Illegal killing of wildlife, both for subsistence and as a commercial enterprise, is an old but still serious problem. Timber harvesting, even when done selectively, often brings large aggregations of laborers, timber camps full of hardworking and famished men and women, who empty a forest of its fauna for meat. Civil and international strife bring hungry, angry soldiers with automatic weapons. War is bad for gorillas and other living things.

None of these concerns is unique to Africa, but given what's at stake, the African particulars deserve special attention from the rest of the world. Africa's glories and successes deserve special attention too. Despite all travails, African peoples produce magnificent art, graceful cultures, terrific music, great works of the mind, and astonishing acts of political and moral courage. Imperialist rhetoric once branded it the "dark continent," but that was blind and stupid, not just wrong. It's bright with variety, tribulation, and joy.

FAY'S INTENT WITH the Megaflyover was to document the ecological dimensions of that variousness. His conceptual starting point was the World Wildlife Fund map of 104 African eco-regions and the Human Footprint project, conceived by Eric W. Sanderson and a team of colleagues at WCS and Columbia University, and articulated in a paper published in the journal *BioScience* in 2002. Sanderson's group used nine different geographic data sets (measuring factors such as road density, railways, population density, nighttime lighting) to represent the weight of human influence all over the planet, including Africa. Fay wanted to cover as many of the 104 regions as time, budget, and politics would allow, collecting an enormous body of data that would reflect incremental gradients between wilderness and urban glare, between stewardship and abuse, between what is possible and what is actually happening on the ground. Then he would present this database to decision-makers and allocators of money—in Africa, in the U.S. Congress, wherever—and say, Here's some information that might be relevant to your resources-and-security planning.

He recruited Ragg, an experienced bush pilot (and, in an earlier life, a successful optometrist in Austria), who offered his flying skills and the use of his two vintage airplanes, one for primary data gathering, one for support. Ragg in turn enlisted his fellow Austrian, Mario Scherer, who had found African bush flying a lively change from his previous work as a war-crimes investigator in Kosovo. Fay drummed up support from various sources—the Human Footprint lab at WCS, the WILD Foundation, the Bateleurs (an Africa-based organization of bush pilots volunteering for conservation), and as chief financial sponsor, the National Geographic Society. The first takeoff was on June 8, 2004, from Air Force Base Swartkop near Pretoria, soon after which—okay, it was five minutes—Fay's network of digital gizmos suffered an outage. The camera quit, the computers went to battery power, and he sniffed a hint of electrical fire. Oh well, he thought, better a data-system meltdown than full-on engine failure within sight of the runway. He rerigged.

Hopping his way across southern Africa and then northward on a sinuous chain of one-day flights, Fay arranged collaborations wherever possible with local conservationists, field scientists, or national agencies, assisting them with their aerial-survey needs as well as adding data to his own comprehensive trove. In Namibia he worked with Keith Leggett, a researcher tracing movement patterns of desert-dwelling elephants. In Tanzania he helped David Moyer and other members of a WCS team, in conjunction with TANAPA, the national parks agency, on an assessment of crucial ecological corridors connecting protected areas. In Chad he partnered with Malachie Dolmia, a young scientist at the Ministry of Environment, to look for populations of Barbary sheep, dorcas gazelles, and other large mammals in areas outside the Chadian national parks. In Kenya he offered flight hours to Iain Douglas-Hamilton, the eminent pachyderm expert and founder of Save the Elephants (see "Family Ties," pages 129–146). In Niger, besides teaming with Ascani and SOS Faune, he got crucial help from consulting conservationist Hubert Planton and coordinated his mission with the Directorate of Fish and Wildlife, through its director, Ali Harouna. Wherever he went, Fay tried to complement the aerial data gathering with contacts, conversations, and observations on the ground. There was so much to learn and, for his purposes, very little that wasn't relevant.

MANY COMPUTER CRASHES, camera shutdowns, and other minor problems followed that first glitch above Swartkop. Most were easily repaired. There were also a few dire aviation scares, caused by high winds, drastic loss of oil pressure, and other forms of mischance. But Fay was persistent—sometimes, in Peter Ragg's candid view, crazily and obnoxiously so. By the time I met them in Niger, Fay and his pilots (accompanied intermittently by Ragg's pilot wife, Hannelore) had flown 600 hours, crisscrossing 16 countries, usually at about 500 feet above the ground. The vertical camera, firing steadily at three shots a minute, had taken about 92,000 images. One of the Cessnas had gotten a new engine. Both planes needed maintenance.

Team spirit was in disrepair too. Tensions had grown. Flying a Cessna four or five hours a day at low elevation, day after day, for six months, gets exhausting. And flying in the Sahara during the period of harmattan, dry easterly winds, is harder on planes and pilots than flying over the plush canopy of a tropical forest. A sandstorm can be deadly. Fine dust (along the desert's southern edge, where there's some dry soil, not just sand) carries subtler dangers. Such dust sometimes rises even on light winds, filling the air at low elevations with a brown haze that erases the horizon, threatening vertigo. A pilot caught in that guck might steer into a hillside or, if he ignored his instruments for a moment too long, lose track of horizontality altogether and slam-dunk himself into the ground. Dusty blur-outs had bedeviled Fay's group in Chad and continued to cause anxiety during the weeks I was with them in Niger. After our first desert flight, Ragg told me, the new air filter he had installed was already filthy. He scooped out a tablespoon's worth of powdered Sahara. Not good. A plane's engine must breathe. And grit in the cylinders—ugh. Plus, what could you see, what could you accomplish? Unless conditions improved, Ragg argued, we should abort the Niger sequence and get out of here. "It's a waste of fuel, it's a waste of time, it's a waste of engine," he said. "And at the end of the day"—that is, if a plane choked and went down—"it's a waste of life."

Fay's view, conveyed to me while we were aloft with Mario, was more sanguine. Flying in bad air was like "living the life of a carp rather than living the life of a trout," he said. "I'd rather be a trout." But if the food was on the bottom, in murky water, he'd be a carp. As for me—at least until my stomach adjusted to the low-altitude scanning, the sepia air, the smeary horizon, the tight circling, and the fumes vented by that auxiliary tank—I felt more like a dyspeptic walleye.

But visibility improved, off and on, and we kept flying. From the air over Niger, we enjoyed some notable sights: a pair of addaxes skittering like sand crabs along the lee of a linear dune, seven Barbary sheep galloping up a kori, sausage-like towers of dark sandstone along the escarpment of the Djado Plateau, camels standing stuporous and serene in the middle of nowhere. Near one village we gawked down at a cluster of saltmaking pits, each pit a nice disk, variously sized, variously colored, shining azure or turquoise or coppery green from the mineral solutions of their individual sumps, all together like a necklace of bright-colored gems.

Mostly what we observed and recorded, though, were variations on the theme of absence: absence of people, roads, wildlife, water, vegetation—absence of topography itself. Sometimes we looked down and saw nothing but beige flatness, just nothing, even when the visibility was good. Some days we flew a 400-mile loop without glimpsing a single animal, and dozens of miles without spotting so much as a plant. Never mind. Even absence is a form of data. Niger is a country of austere landscapes, spectacularly desolate in their own ways but further desolated by recent human-caused losses. The addax is nearly extinct here, for instance, and the Barbary sheep, and the desert cheetah; their disappearance from remote habitat areas may correlate with the presence of four-by-four tracks, indicating unimpeded access by poachers. Such tracks show clearly from 500 feet up. Measures could be taken. Fay would offer all his findings to the Nigerien officials responsible for protecting what remained.

ONE THING THAT remained, tangible on the ground and indelible in my memory, was the assemblage of dunes at Arakao. Disappointed at having seen no sheep in the Aïr Massif, suddenly we were thrilled. We all pressed to the windows, even Ascani, the old Niger hand. Then, as we circled wide around those great soft mounds to admire them from all angles, another odd sight came into view: a large green oval. It was a pond, evidently spring-fed from beneath the sands. Water?

Fay peered down for a moment and then, having noticed something, tapped a note into his tablet: "no animal tracks." It hadn't struck me, but of course: A water hole out here should attract gazelles and other mammals from many miles around—attract them, that is, if any exist. He tapped again: "4x4." Meaning, tire marks from an overland vehicle. An absence of animal sign, a presence of human sign. Cause and effect? Anyway, data.

I was still mesmerized by the dunes. They seemed the perfect icon for what Mike Fay had strived to produce, first in his Megatransect, and now by way of his Megaflyover: a vast aggregation of tiny particles, in the form of data, heaped together for larger effect. The individual grains were innumerable and insignificant. The task of collecting them was tedious, arduous, and risky. It couldn't be done by just anyone. It took a dry mind and a will like desert wind.

The tricky part that we had to remember, and he too, was that Arakao is more than a huge sandpile. It's an uncannily beautiful, huge sandpile.

The last challenge for Fay, as he processed and presented his data, was parallel to that miracle of beauty, but slightly different. From a mountainous pile of facts and photographs he had to deliver meaning.

. . . .

One of the ironies of the Megaflyover, viewed in retrospect almost 20 years later, is that it proved more important to Fay and his future conservation efforts for what it didn't allow him to see than for what it did. It didn't allow him to see southern Sudan. It didn't allow him to gaze down and photograph the woodlands and clay plains sprawling east of the White Nile River, north of the city of Juba, and a vast wetland contained by a bend of the great river and known as the Sudd. It didn't give him a glimpse of the extraordinary riches of wildlife living in those places, and migrating between them, including almost a million individuals of a species of antelope called the white-eared kob.

"During the Megaflyover, we never made it to South Sudan," Fay told me recently. The reason they never made it was that South Sudan, as a country, didn't yet exist. Rebel forces of the Sudan People's Liberation Army were fighting in a revolt against the Sudanese government based up north in Khartoum that had already lasted almost two decades. "We didn't get there because it was still highly insecure and you needed special authorization to get in there, and there were sanctions and all kind of stuff."

"The war was still on," I said.

"The war was still on," he confirmed.

Then it ended, with a peace agreement in 2005, six years of provisional autonomy for southern Sudan, and a referendum that yielded a landslide vote for independence. The Republic of South Sudan came into existence on July 9, 2011.

In the meantime, Fay began planning another aerial survey, to focus specifically on lands along the eastern half of that transitional *zone south of the Sahara and known as the Sahel. By then he was working for WCS in Zakouma National Park, in southeastern Chad. He wrote a proposal for an ambitious, expensive enterprise he called the Sudano-Sahelian Initiative, which would span 10 countries across Africa's middle, from west to east: Mali, Niger, Burkina Faso, Benin, Ivory Coast, the Democratic Republic of the Congo (DRC), the Central African Republic (CAR), Chad, southern Sudan, and Ethiopia. It would encompass two dozen jeopardized and biologically rich landscapes, and it would cost, he figured, about $125 million a year. This was typical of Fay's convictions, his disposition, and his approach: Earth is losing biological diversity at a catastrophic rate, so think big, act boldly, be tough on yourself and on others to meet the challenge—or go home.*

In 2005, he did an aerial survey of northern CAR, part of that Sudano-Sahelian band. He tried again, unsuccessfully, to get into southern Sudan, but the war and the sanctions still made that impossible. "We kept lobbying and pushing on it, and it kept going back to the State Department and to DOD and everybody else," he said, referring to the U.S. Department of Defense, which suggests the delicacy of overflying that area, "and finally I was able to get the authorization." He flew above the clay pans and wetlands and migration corridors, taking photos and making systematized counts, and found "almost a million kob and hundreds of thousands of other antelope," such as tiang, a subspecies of topi, a robust, reddish brown cousin of the hartebeest. "It was one of those life-changing experiences," Fay told me. "I mean, that is the first and last time I've ever seen hundreds of thousands of animals"—big animals, stampeding hordes, some fleet, some lumbering, and not merely that but "hundreds of thousands of animals that no one knew were there."

He parted ways with his old employer, the Wildlife Conservation Society. "I jumped ship to African Parks," he told me. (See "Boots on the Ground," about the African Parks organization, pages 309–324.) "They understand the local people, and they get the job done," he said. I had heard of that switch several years ago, through the conservation grapevine. Where's Mike Fay? Is he still in Gabon? Is he back at that cabin in Alaska? Is he walking across the Amazon? No, he's in CAR, working for African Parks, serving as interim director general of Chinko National Park.

Three months ago, the grapevine spoke again: No, Fay's not in CAR. He's now in South Sudan, scoping the possibility of a major role for African Parks in managing Bandingilo and Boma National Parks, and the migration corridors between them, and possibly two other parks across the border in Ethiopia. "It's kind of the last great frontier that's not managed in that whole Sudano-Sahelian concept," he told me when I caught up with him on a tenuous WhatsApp connection. "If we don't do it now, and it's not done by somebody competently, it's never going to happen."

I knew he was busy, and that it was evening in eastern Africa. "Stay healthy. Stay safe," I told him as I ended the call.

"Okay."

"Keep doing what you're doing."

"Yeah," he said. "See you somewhere, sometime."

THE SPIRIT OF
THE WILD

·❦·❦·❦·

Africa Has Saved What Other
Continents Have Lost

The September 2005 National Geographic *was a special issue devoted entirely to Africa. It contained some keen and important feature stories by some masterful writers and photographers that I've been proud to call colleagues: the writer Paul Salopek and photographers Lynn Johnson, David Alan Harvey, and Randy Olson, among others. The issue illuminated a sampling of the peoples, cultures, and issues of what is arguably Earth's most multifarious continent. My part of the effort entailed two different pieces, both included here. I contributed the story on Mike Fay's Megaflyover (see the previous chapter, pages 109–122), in partnership with the dauntless airborne photographer George Steinmetz. And I was asked by Editor in Chief Chris Johns to write a short essay to serve as a sort of coda to the issue. It would accompany a portfolio of a dozen or so powerful photos of African wildlife, including two by Nick Nichols and one by Chris himself from his years as a staff photographer. Pondering those photos got me thinking about something that hadn't occurred to me before but became the point of the piece. Please bear in mind: This was just a small verbal bookend, intended to read as a reflection, not just on a portfolio of photos but on all the diverse reporting and luminous photography that preceded it in that issue.*

S o, Africa is complicated. The continent's most pressing concerns, yes, are political and economic and medical. Landscape itself is often a battleground here, not just for armies but also for opposing visions of resource exploitation, conservation, and governance. Africa is fraught with issues that demand careful study, cool discussion, hard choices, compromise, meticulous attention to boring details, and patient planning; that call upon a range of human skills, including diplomacy and sociology. But beyond all the complexities, at the end of the day, at the start of our new millennium, there's another salient fact: Africa is an extraordinary repository of wildlife. It's the greatest of places for great beasts.

This fact, which seems so simple that people take it for granted, is a bit complicated itself. The roster of species, for starters, is dizzyingly diverse: three big cats (lion, leopard, cheetah), seven smaller cats (such the caracal and the serval), two species of elephant (savanna and forest), two rhinos (black and white), two hippos (pygmy and regular), four species of non-human ape (two gorillas, the common chimp, and the bonobo), three zebras, nine gazelle species, 19 duikers, dozens of monkeys, five species of baboon, a gaggle of genets and civets, six different pigs, four pangolins, three reedbucks, some horsey antelopes, some dwarf antelopes, nine species of spiral-horned antelope (including the bongo, sitatunga, and eland), two wildebeests (that's gnus to you), the aardvark, the aardwolf, the drill and mandrill, the rhebok, blesbok, gemsbok, wopbopaloobop (only kidding), the African buffalo, the Nubian ibex, three hyenas, three jackals, the Ethiopian wolf, the wild dog, and many other mammals, not to mention two species of ostrich, three species of crocodile, the African rock python, the ball python, plus sharks and other sizable fish in the coastal waters, as well as smaller terrestrial animals of every imaginable sort. Wow. It's a spectacular assemblage, both in variety and abundance, unmatched elsewhere in the contemporary world. But to appreciate fully what's present in Africa, you need to consider what's absent elsewhere, and why.

That's the task of paleontologists, who study living creatures no longer alive—the biota of the past. Their data come from the fossil record, and their vast backward calendar of Earth's history is demarcated by episodes of mass extinction. Each such episode represents an abrupt loss of biological diversity and marks a boundary between two (otherwise arbitrary) units of time. By the end of the Cretaceous period, for instance, 65 million years ago, there were no more surviving dinosaurs—not by coincidence, but because the disappearance of the dinosaurs is one of several important factors *defining* the end of the Cretaceous. At the close of the Permian period, 245 million years ago, came another massive die-off, cataclysmic and sudden, eliminating about 95 percent of all animal species in existence then. The Pleistocene epoch, a more recent unit that ended just 10,000 years ago, is also known for its extinctions, especially among big mammals and huge, flightless birds. The mammoths and mastodons vanished, along with giant sloths, giant bears, giant beavers, saber-toothed cats, giant kangaroos, and countless other large-bodied animals. Many of those Pleistocene extinctions occurred near the end of the epoch, most notably in North America, South America, Australia, New Zealand, and Madagascar. What caused them? Nobody knows for sure. Lethal changes of some sort, still mysterious, still debated by experts, affected those continents and islands. Probably the arrival of humans was at least part of the problem. We appeared from nowhere, armed and dangerous and hungry.

Africa was different. Africa suffered only modest losses of fauna during the Pleistocene (which began about 1.8 million years ago) and no widespread, severe, or simultaneous set of extinctions at its close. Mostly the large African mammals of 20,000 years ago have survived as the large African mammals of today. For that reason, Africa has been called the "living Pleistocene." It stands to remind us of an epoch, before the rise of *Homo sapiens,* when the planet was *really* big and wild.

But remember another thing. The survival of Africa's wildlife hasn't depended on an absence of people. On the contrary, it has happened amid constant human presence. We ourselves belong to a species of African origin, *Homo sapiens.* We first emerged in this place, attaining our current shape, brain size, social instincts, and sense of identity during millennia spent as members of rough-and-tumble African ecosystems. The other animals adjusted to our presence—to our slowly but radically increasing capabilities—even as we adapted to life among them. One lesson along the route to civilization, learned accidentally by African peoples, and evidently not portable when humans dispersed elsewhere, was the possibility and rightness of coexisting with other formidable creatures, even those sometimes as menacing as ourselves.

That was a virtue derived from necessity. And now the necessity is gone. Killing wildlife, extinguishing species, and destroying habitat are easy with our current weapons and tools. Preserving the last of the great beasts in their landscapes, despite human needs and pressures round about, is more difficult. But wait, here's a thought, unabashedly hopeful and as wild as an aardvark: Maybe modern Africa is where we can rediscover how it's done.

....

Now a coda to this coda: I would never forget, nor wish you to forget, that the peoples and the landscapes of Africa are beset with a vast number of challenges in addition to the challenge of conserving their native wildlife; that they are distinguished by a vast number of glories and beauties and surprises and conundrums beyond the glorious biological diversity that remains there. I'm bemused sometimes—and I try to remain politely speechless—when I'm introduced to someone who hears I've lately been working or traveling "in Africa," and they cheerily offer: "Africa! We love Africa too! We went there again just last year." Usually this comes from an affluent American, and most often what they're talking about is a vacation watching wildlife in Kenya and Tanzania. Don't get me wrong: Watching wildlife in Kenya or Tanzania is a wonderful, soul-opening activity, and I recommend the experience to anyone in the world who can afford the ticket. But even a location so magnificently African as, say, Kenya's Amboseli National Park is not to be equated with "Africa." Africa is a spectacular quilt of landscapes and cultures—as I mentioned in "The Megaflyover" (pages 109–122): 54 countries, hun-

dreds of ethnic groups, and . . . ah yes, the only four species of giraffe currently extant on planet Earth (plus that giraffian cousin, the okapi). It's a blessed privilege to be able to see them there, still, in the 21st century. The entire world owes a debt of gratitude for that, as for so much else, to Africa: the miracle of the giraffe.

FAMILY TIES

·┤·┤·┤·

The Private Lives
of Samburu Elephants

....

"Happy families are all alike," wrote Leo Tolstoy; "every unhappy family is unhappy in its own way." But the great Russian novelist wasn't thinking of elephants, whose familial relations are complex, various, and fascinating, even amid the relatively happy circumstances of Kenya's Samburu National Reserve.

....

The biologist Iain Douglas-Hamilton was walking up on an elephant, a sizable young female, nubile and shy. Her name, as known to him and his colleagues, was Anne. She stood half-concealed within a cluster of trees on the knob of a hill in remote northern Kenya, browsing tranquilly with several members of her family. Around her neck hung a stout leather collar along which, at the crest of her shoulders, like a tiny porkpie hat, sat an electronic transmitter. That transmitter had allowed Douglas-Hamilton, flying in by Cessna, proceeding here on foot through the tall grass and acacia scrub, to find her. Crouching, he approached her, moving upwind to within 30 yards. Anne gobbled some more leaves. She was oblivious to him, or maybe just not interested.

Elephants can be dangerous animals. They are excitable, complex, and sometimes violently defensive. Douglas-Hamilton is a world-renowned expert who has studied them for 40 years. Don't try this at home.

He wanted a clear look at the collar. He had heard reports that it might be too tight—that she had grown into it since having been tranquilizer darted, fitted, and thus recruited as a source of research data. Ordinarily, Douglas-Hamilton did his elephant-watching more cautiously, from the safe containment of a Land Cruiser, but no vehicle could drive this terrain, and Anne's comfort and health were at issue. The collar was meant to hang loose, with a dangling counterweight below. He wanted to be sure that Anne's wasn't snugged up to her throat like a noose. But here, amid the thicket, she was showing him only her imperious elephantine butt. So he crept closer.

Three other men lagged back. One was David Daballen, a bright young Samburu protégé of Douglas-Hamilton's, who often accompanied the boss on missions like this. The second man was a local guide holding a Winchester .308 rifle. I was the third. As we watched Douglas-Hamilton edging forward, we noticed another female elephant, a big one, probably the group's matriarch, sidling around craftily on his right flank. We ducked low to escape the matriarch's view. We froze. As this female came on, suspicious and challenging, Douglas-Hamilton seemed unconcerned with her, but Daballen began to look nervous. He was calculating (he told me later) how fast an elephant might be able to charge across such a rocky, rubble-strewn slope.

Then the big female committed herself to a sequence of gestures suggesting nonchalant disregard: She pissed torrentially, she defecated galumphingly, and she turned away.

Anne herself swung daintily out of the brush. She stepped toward Douglas-Hamilton. The gap between them was 50 feet. For a few seconds the young female graced him with a frontal view of her large forehead, her flappy ears, her pretty tusks, as though posing for beauty shots in the glow of a flash. She gave him a profile. He raised his camera and clicked off several frames. Then she too turned and ambled away. Through his lens, in those seconds, he saw that the collar was hanging just as it should. The alarm was a false one. Anne was in no danger—or anyway, no danger of chafing or choking.

As we withdrew, circling back toward our vehicle, I thought, So that's how it's done. Show a little caution, a little respect, get the information you need, back off. And everybody is happy. After four decades Douglas-Hamilton knew this species about as well as anyone in Africa. He had a keen sense, well earned by field study and sharpened by love, of the individuality of the animals—their volatile moods, their subtle signals, their range of personalities and impulses. Nothing about this interaction with Anne prepared me for the moment, some weeks later, when I would watch him charged, caught, thrown, and nearly tusked through the gut by an elephant.

SOON WE WERE aloft again in Douglas-Hamilton's Cessna, flying low over the contours of the landscape. It was his preferred style, flying low; why go up a thousand feet when you could float just above the terrain, admiring it more intimately? So we rose and descended gently over the rocky slopes, ridges, dry acacia plains, sand rivers, returning northeast toward a place called Samburu National Reserve. Just beyond the reserve sat a gravel airstrip and, not far from that, his field camp. We would be home before dark.

Samburu National Reserve is one of the little-known jewels of northern Kenya, taking its name from the proud tribe of warriors and pastoralists in which David Daballen, among others, has his roots. The reserve is a relatively small area, just 65 square miles of semiarid savanna, rough highlands, dry washes (known locally as *luggas*), and riparian forests of acacia and doom palm along the north bank of the Ewaso Ngiro River. Lacking paved roads, sparsely surrounded by Samburu herders, it teems with wildlife. There are lions, leopards, and cheetahs, of course, but also Grevy's zebras, reticulated giraffes, beisa oryx, gerenuks, Somali ostriches, kori bustards, and a high diversity of showy smaller birds such as wattled starlings, pin-tailed whydahs, and lilac-breasted rollers. But the dominant creatures are the elephants. They play a major role in shaping the ecosystem itself—stripping bark from trees or uprooting them, keeping the savanna open. They intimidate even the lions. They come and go across the boundaries of the reserve, using it as a haven from human-related dangers in a much larger and more ambivalent landscape.

The larger landscape includes all of Samburu County (within which the reserve lies) and parts of three other counties, most notably Laikipia, a high-elevation patchwork of private ranches and sanctuaries, community conservation areas, wheat fields, fences, mountain slopes, stream valleys, roads, and *shambas* (small family farms) just to the south. In Laikipia, zones of wildlife habitat, crop production, cattle husbandry, and human habitation are juxtaposed like a spilled box of multicolored mosaic tiles. Samburu, by contrast, has fewer shambas and scarcely any fences. The Samburu people, who speak a dialect of the Maa language, have shown little inclination to surrender their traditional ways—tending goats and cattle, costuming themselves resplendently (especially the young men) in beads and feathers and red *shukas* (blankets), exchanging raids against their ancient enemies—in favor of modern, pusillanimous practices such as growing crops. Their traditionalism, along with a shortage of good soils and water and a growing awareness of the economic benefits of tourism, had so far spared Samburu County from the sort of intensive land conversion seen in parts of Laikipia. The combined Samburu-Laikipia ecosystem comprises roughly 11,000 square miles, and within it, at the time of my first visit in 2007, lived about 5,400 elephants—the largest population of *Loxodonta africana* existing mainly outside protected areas anywhere in Kenya.

That population size, and its considerable growth over the preceding 15 years, reflected the fact that Samburu-Laikipia is a productive, hospitable landscape for elephants, but two other adjectives are also applicable: edgy and complex. Within the mosaic of mixed uses and shifting seasonal conditions, elephants face certain risks. So do people. Conflicts occur, resulting occasionally in a crop devastated by raiding elephants, or a cow killed, or an elephant shot, or a person trampled and tusked. And with Kenya's human population also growing by more than 2 percent annually, the potential for such conflicts could only increase. Decisions would be made about what to protect (elephant travel corridors? cornfields? the right of people to continue establishing new farms?) and what would be sacrificed. Douglas-Hamilton's goal was to supply the makers of those decisions with scientific information more detailed and more timely, and therefore more useful, than any hitherto available. It wasn't precisely the same research agenda with which he began his long career, but it shared the same spirit and the same guiding concerns. It was where the contours of the landscape had led him.

"If you had asked me, when I was 10 years old, what I wanted to do," Douglas-Hamilton told me, "I'd have said: I want to have an airplane; I want to fly around Africa and save the animals."

Aviation was part of his lineage. His father, Lord David Douglas-Hamilton, commanded a Spitfire squadron in the Battle of Malta and then died on a reconnaissance mission later in World War II. His three uncles were also distinguished Royal Air Force (RAF) fliers. One of those uncles, before the war, was the first man to pilot an open-cockpit biplane over the summit of Mount Everest, just for the sheer glorious hell of doing it. (He dressed warmly before climbing in.) After the war, Iain's mother remarried, this husband a kindly man who read Iain stories about Africa and who took the family to live in Cape Town, then died abruptly himself. At age 13, Iain found himself back in Britain at a Scottish boarding school, nurturing dreams of a getaway. As an undergraduate at Oxford, he would have joined the RAF Volunteer Reserve, following the path of his father and uncles, but poor vision disqualified him. Zoology, fortunately, didn't require 20/20 eyesight.

"Science for me was a passport to the bush," he said, "not the other way around. I became a scientist so I could live a life in Africa and be in the bush." Almost wistfully, he added, "I would've liked to have been a game warden." But for a young Scotsman who spoke no Swahili, in the early 1960s, just before Kenyan independence, such a civil employment position was out. Instead, he went to Tanzania as a research volunteer and then was offered a project in a small area called Lake Manyara National Park. With a bit of money from selling some inherited stock, he bought himself an old 150-horsepower Piper Pacer, nimble enough for tracking big animals, and learned by trial and error to land it on rocky airstrips.

There at Manyara, Douglas-Hamilton did the first serious study of elephant social structure and spatial behavior (where they go, how long they stay) using radio telemetry. It earned him a doctorate at Oxford. He also became the first student of elephants to focus closely on living individuals, not just trends within populations or the analysis of dead specimens. He used photographic records of visual patterns—unique ear notches and perforations, tusk shapes—for identification of animals in the field.

He got to know the elephants one by one, noted their individuating traits, gave them names, watched their social interactions. He had a favorite named Boadicea, a great matriarch with long tusks that converged almost to a point, who made emphatic threat charges but whose bluff could be called by standing firm. There was another, a one-tusked female he called Virgo, very different from Boadicea, who acquired the habit of approaching his vehicle and reaching out toward Douglas-Hamilton with her trunk. After four years of slowly decreasing wariness, she would greet him with raised trunk and let him tickle her on its sensitive underside. He witnessed the infancy of a male named N'Dume, born to a female called Slender Tusks; he watched the calf learn to suckle, to use his trunk efficiently for grazing, and (on pain of chastisement) to avoid collapsing the water holes his mother had dug. Noticing the distinct traits of individuals and the generalized patterns within a population, Douglas-Hamilton began to wonder about motivations. What did elephants need? What did they want? How did their movements on the landscape reflect those cravings? What sort of choices did they make?

He married a Kenyan-born Italian woman named Oria Rocco and took her back to the Tanzanian bush, where she shared his field life and passion for elephants. Together, during the 1970s, they produced one best-selling book, *Among the Elephants,* and two luminous daughters. Photos from the time show Iain Douglas-Hamilton as a thin young man with wild, sun-bleached hair and nerdy glasses, wearing bush shorts and boots, sometimes a field vest but no shirt, deeply tanned, living a dashing life amid friendly pachyderms: an amalgam of Tarzan, Clark Kent, and Doctor Dolittle.

Then came the grim years of the late 1970s and '80s, when Douglas-Hamilton played a lead role in raising the alarm against an ugly development—the wholesale slaughter of African elephants. Killing elephants for their tusks wasn't new, of course. People have been doing that ever since the invention of the spear. But this modern phase, driven by a sudden sharp rise in the price of ivory and made gruesomely efficient by automatic weapons, was on a different scale. Between 1970 and 1977, according to one assessment, Kenya lost more than half its 120,000 elephants. Ivory exports from the continent—just the legal exports to major markets, not even considering small markets or smuggling—totaled about two million pounds a year. Based on that weight of tusks, Douglas-Hamilton calculated elephant losses throughout Africa at somewhere above 100,000 animals annually. He decided to do something.

With funding from several conservation NGOs, he organized a hugely ambitious survey to gauge the status of elephant populations throughout the continent. He mailed out questionnaires to field biologists, game wardens, conservationists, and other well-informed people, asking for their counts or best estimates of local and regional populations, and he flew surveys himself. From the results, compiled in 1979, he figured that Africa then contained about 1.3 million elephants. It might seem like a sizable number, but there was a devil in the details; the trend lines pointed down. African elephants were dying at an unsustainable rate, Douglas-Hamilton concluded, putting the viability of their populations at risk.

Some experts in the field disagreed, arguing that elephant populations were doing just fine, or at least that Douglas-Hamilton's data were unreliable. Those disagreements eventually carried through the 1980s in a series of contentious meetings and bureaucratic battles that became known as the Ivory Wars. Douglas-Hamilton had meanwhile set aside his behavioral studies and spent years investigating the status of beleaguered elephant populations in Zaire, South Africa, Gabon, and elsewhere, both by overflying to count animals and by amateur sleuthing on the ground. He went to the Central African Empire, nosed into the ivory trade there, and left quickly when the emperor, Jean-Bedél Bokassa, began to get curious about this visiting elephantologist. He flew into Uganda amid the turmoil after Idi Amin's fall and saw bullet-riddled elephant carcasses littering the national parks.

"It was a dreadful time. I really spent a terrible 20 years doing that," he told me. His dangerous, gloomy work helped immeasurably to support the 1989 decision under the Convention on International Trade in Endangered Species of Wild Fauna and Flora (CITES) to outlaw the international sale of ivory. But on a personal level, it cost him—anxiety, years of his life, time away from his wife and daughters, and time away from living elephants.

One of his colleagues in Kenya, Cynthia Moss, herself a highly respected elephant behavioralist, called my attention to that last category of cost. It's so important for an elephant researcher, she said, during a lunch in Nairobi, to be in the presence of known individual animals. Flying is useful, counting is useful, but those are no substitute for close, prolonged observation. Moss began her own career, back in 1968, as a research assistant to Douglas-Hamilton at Manyara, and she voiced an old friend's sympathetic concern. "He didn't really come back to ground," she told me, "until he started up in Samburu."

His work in Samburu National Reserve reflected a new role in Douglas-Hamilton's life: mentoring young scientists. He came, in 1997, along with a student he had placed there.

The student was George Wittemyer, an American Fulbright scholar who wanted to study elephant social relations. By that time, Douglas-Hamilton had established his own research and conservation organization, Save the Elephants (STE), based in Nairobi. He supplied Wittemyer with contacts, an aegis, and a couple of used tents, with which Wittemyer set up a simple field camp along the Ewaso Ngiro River, in the shade of some large acacia trees and near a conical hill. Just as Douglas-Hamilton had done three decades earlier at Manyara, Wittemyer began learning the local elephants, sorting out their family affiliations, and naming them.

As in other elephant populations, each family was dominated and guided by a matriarch, an older female, who was mother or grandmother to most of the members. Wittemyer grouped the names in mnemonic familial clusters, a system later researchers at Samburu continued: the Royals (Victoria, Cleopatra, Anastasia, Diana), the Biblical Towns (Babylon, Nazareth, Jerusalem), the First Ladies (Eleanor, Martha, Lucy Kibaki, Jackie), and many others. Bulls tend to travel solitarily or in male affiliations, so the Samburu bulls were named more variously: Mungu (Swahili for God), Gorbachev, Mountain Bull, Genghis Khan, Marley, Amadeus, etc. Roughly 900 individual elephants used the Samburu reserve in the course of a year, either as residents or as short-term visitors, and most of them were identified in STE records.

Gradually the two-tent camp became a permanent compound, ascetic but comfortable, comprising a dozen wall tents, a thatch-roofed kitchen, an office and dining hall structure with a concrete floor and wireless internet, plus outhouses and bucket showers. Douglas-Hamilton made this gracious outpost, dubbed simply Save the Elephants Camp, his base of operations.

The project grew—into a doctorate for Wittemyer and a long-term monitoring program on social and spatial behavior for Save the Elephants. Other young men and women turned up, from within Kenya and far beyond, and with guidance from Douglas-Hamilton, assumed a variety of responsibilities. Onesmas Kahindi, Maasai by descent but Samburu by affinity, took over the behavioral study, then found a role better suited to his gifts: gathering data on elephant mortality. Tall and charming, a natural schmoozer, Kahindi prowled the ecosystem like a traveling salesman, using records from the Kenya Wildlife Service (KWS) and his own network of local informants to guide him to every elephant carcass—both natural mortalities and elephanticides—that turned up. By documenting and tallying all those deaths, he maintained a crucial detection system (part of an international program called MIKE, meaning Monitoring the Illegal Killing of Elephants) against resurgent poaching. Henrik Rasmussen, an ecologist from Denmark, complemented Wittemyer's work on female behavior with a study of reproductive tactics among the males. David Daballen, my companion for the outing to see Anne, was a high school graduate with a Ph.D. mind, recruited by Kahindi from a group of volunteer rangers; he worked as a field assistant until Rasmussen recognized the greater scope of his potential. Daballen became camp manager as well as co-researcher on the long-term behavioral study. Daniel Lentipo, another local Samburu, with keen eyes and a wizard memory, became the other chief research assistant on that study. Douglas-Hamilton said: "I love the interface between these high-powered, overseas scientists and the Samburus of camp."

Of the 900 elephants inhabiting or passing intermittently through Samburu National Reserve, Daballen and Lentipo could each recognize about 500 individuals on sight. Having watched births, matings, deaths, and group behavior over time, they also knew the family histories. Daballen could tell you, for instance, that this magnificent female on the south bank of the river, so huge she looked like a bull, was Babylon, matriarch of the Biblical Towns; that she was nearly 50 years old; and that her breasts were full because that was her young calf standing nearby, along with her older daughter and granddaughter. He could point out a youngish female limping piteously on three legs and identify her as Babel, of the same family, probably crippled when she was mounted too young by a bull, and that the other Biblical Towns, taking their cue from old Babylon, moved slowly so that Babel could stay with them. When Daballen looked at an elephant in this ecosystem, he saw an individual with a story embedded in a matrix of relationships and other stories.

Meanwhile Douglas-Hamilton, working with still other young collaborators, concentrated on the spatial dimension—that is, the study of which elephants move where, and when—using GPS technology. The spatial study formed a high-tech complement to the low-tech behavioral observations.

DOUGLAS-HAMILTON COULD RECALL the first GPS unit he ever saw in action, brought to Kenya by friends in 1991 and rigged onto an airplane for use in counting elephants within Tsavo National Park. That unit registered only the whereabouts of the airplane, as the plane traced the whereabouts of the animals. Still, he said, "It was quite a revelation to see how the elephants moved and circled—the patterns they adopted." The patterns were important because they reflected informed choices by the elephants as to where they might best satisfy their most urgent imperatives, finding what Douglas-Hamilton called the three S's: sex, sustenance, and security. GPS offered a way to chart such patterns in detail. "As soon as I saw one of those gadgets," he told me, "I wanted to put it on an elephant."

About 20 elephants in the Samburu-Laikipia ecosystem, at that stage of the work, wore Save the Elephants GPS collars. The latest model as of then (which might seem primitive to us now) delivered one positional point from each elephant every hour. The technology of the collars was not only intricate but also economical: To save expense, conserve battery power, and minimize weight, those units received positional information from GPS satellites but then transmitted that information by way of low-cost SMS (short message service) blurts on Safaricom, Kenya's leading cell phone network. In other words, everybody in Kenya had a mobile phone by 2007, even the elephants. A few Grevy's zebras were also transmitting via Safaricom, and one could only imagine what might be next—rock pythons, kori bustards, lilac-breasted rollers? For the time being, what this meant was that each of 20 elephants sent a text message to Douglas-Hamilton's computer, each hour of every day, saying: "Iain, yo, here I am."

Save the Elephants had GPS tracking projects under way not just in Kenya but also in Mali, South Africa, and the Democratic Republic of the Congo. One important discovery to come from GPS tracking was what Douglas-Hamilton called "streaking" behavior: the occasional event wherein an elephant or a group of elephants sets out at high speed and travels a long distance in a short time, from one secure area to another, by way of a perilous or at least inhospitable route. A bull known as Shadrack made such a streak—from the green highlands of the Marsabit massif through a town, across the Kaisut Desert, to the Matthews Range of north-central Kenya—covering 50 miles in 36 hours. Another elephant, a female known as Mrs. Kamau, made an even more ambitious streak from Marsabit northeastward, roughly a hundred miles in 48 hours, to a solitary zone of lava-paved desert, where she somehow found water and food as well as security in a sere, ragged landscape. Still another male, Mountain Bull, performed an astonishingly discreet series of down-and-back streaks, traveling between the safe northern slopes of Mount Kenya, through a maze of villages, wheat fields, roads, and a safe area along a Laikipia canyon— making this journey not once but 14 times within the space of a year. Each of these animals was a collared representative of what was possibly a whole group of streaking elephants. Their cross-country dashes, recorded by Douglas-Hamilton's system, and interpreted by him collaboratively with a senior scientific colleague, Fritz Vollrath of Oxford University, helped delineate crucial travel corridors within the Samburu-Laikipia ecosystem.

The data points accumulated—about 1.5 million of them, by that time, including both the widely spaced dots reflecting high-speed travel and the less dramatic speckles representing smaller-scale, quotidian movements. Software created by another young team member, Jake Wall, allowed those data to be mapped and animated on Google Earth's vivid topographic Kenya. So if you were Douglas-Hamilton, or anyone otherwise privy to the access codes, you could turn on your computer any morning and see which of the collared animals had gone where. (But that access to real-time data was closely guarded, lest poachers abuse it to locate elephants.) You might have seen that Mountain Bull had streaked back to Mount Kenya; or that Jerusalem, presumably accompanied by her five-year-old calf and probably other females of the Biblical Towns, had descended from the safe hillsides south of Samburu and come to water along the Ewaso Ngiro.

And there was uptake by the decision-makers. When I called on the director of the Kenya Wildlife Service, Julius Kipng'etich, at his office in Nairobi, I noticed two maps of Kenya on his wall. One was festooned with blue pins: antipoaching squads, he said. The other map was crisscrossed with squiggly lines, each line bearing a red directional arrow. "All these red arrows are elephant corridors," the director told me, then added that such data allowed him to present good wildlife-management and land-protection advice to the government. As he spoke, I noticed a single red line running northeast from Marsabit far into the desert, and I thought, There goes Mrs. Kamau.

Of the cardinal incentives that drive elephant behavior—that is, Douglas-Hamilton's three S's: sex, sustenance, and security—the most difficult to calibrate is the third. Finding food, water, and reproductive opportunities aren't always simple tasks, but compared with finding security they are relatively straightforward. Real security, lasting security, is more unpredictable and elusive. The local people have a word for it: *neebei*. Every person wants neebei—freedom from danger, menace, uncertainty, fear—and it's not anthropomorphism to say that every elephant does too.

Even in northern Kenya, even in the 21st century, with the ivory trade banned and the KWS policing against poachers, the life of an elephant (especially a bull with large tusks) can be precarious. Sometimes an animal is killed by an angry farmer who has seen a crop wrecked, or by an outraged herder who has found a precious cow fatally gored and taken vengeance on the next elephant to appear. Sometimes people still kill for ivory, blasting an elephant full of high-caliber slugs, hacking the face off to wrench out the tusks, moving that ivory into the black market. And sometimes an elephant dies an untimely death for reasons that can't be discovered. When I returned to Kenya for a second visit, after a month away, Douglas-Hamilton told me that Anne, the young female whose collar we inspected on foot, was dead.

She had been shot by persons unknown for reasons unknown. Her tusks, smallish but valuable, hadn't been taken; they were still in place when a KWS patrol found the carcass, by which point they could be pulled from the rotting skull without the need to hack. There was no trace of a perpetrator and no clue to a motive.

A week later I visited Anne's remains, this time with Onesmas Kahindi, the carcass-data man. We found her (with help from Iain, who had flown over the last GPS position and caught a glimpse of white bone) in a soggy, spring-seep valley of western Laikipia, just upstream from a rectangular lake. A cruising vulture lingered nearby, but there wasn't much left to interest it.

Anne's skull, resting beneath a yellow fever tree, was painted with bird dung. Her lower jaw, several ribs, a shoulder blade, and other bone fragments lay scattered about, along with a smear of grassy stomach contents and a patch of dried skin. Her jaw joints showed gnaw marks from a hyena. The whole area reeked of death, but approaching downwind, we hadn't smelled her until we saw her. She had been dead about three weeks. The maggots and flies, like the hyenas, had already come and gone. Kahindi measured a molar. He snapped a photo. The sky began darkening toward an afternoon shower as he recorded his data.

Anne had made her choices, and one choice brought her to this little valley, probably for its water and good grass. Whatever else she found, she hadn't found neebei, but the details of her misfortune were inscrutable. Kahindi, a tireless worker for elephant conservation but no sentimentalist, capped his pen. "Finished," he said. "Okay, let's beat the rain."

It's ALL ABOUT choices. Elephants are smart, they know what they need, and they generally know where to get it; if they don't know, their mother or grandmother will teach them. They seem to calculate risks. They can be dangerous, but they prefer to avoid conflict with other big, dangerous creatures such as lions or people. They are herbivores, after all, with no reasons to kill except defense, confusion, panic, and desperation when their needs are unmet. In the Samburu-Laikipia ecosystem they manage to live in the spaces between human farms and settlements with far lower levels of conflict and higher levels of mutual tolerance than exist in most other areas where elephants range. Douglas-Hamilton talked thoughtfully to me about such things, both before and after the day I nearly got him killed by an elephant.

It happened like this: Late one afternoon, he stopped by my tent and asked, Want to drive out and see some elephants before sunset? He often rewarded himself that way for eight hours' deskbound effort. On this occasion I said, How about a walk instead? I knew that foot travel within the reserve was generally inadvisable, but couldn't we at least climb the little conical hill just behind camp? Yes indeed, he said; and so we did. From the hill's rocky top, we savored a magnificent view westward, with the brown slick of the Ewaso Ngiro winding its way downstream between banks bristling with palm and acacia. Just north of us was a larger hill, a double mound known as Sleeping Elephant. Have you ever climbed that one? I asked. No, said Douglas-Hamilton, with a mischievous glint in his eye . . . but we could.

Thus we set out on foot toward Sleeping Elephant: two middle-aged white men and a young Samburu from the camp crew, a skinny lad named Mwaniki, in his beads and shuka, whom Douglas-Hamilton asked to tag along. We walked only five minutes through the high, sparse brush before we saw elephants ahead: a female with two calves. We paused, admiring them from a safe distance until they seemed to withdraw, and then we went on. Seconds later Mwaniki muttered a warning, and we looked up to see the female glowering at us from 70 yards away. Her ears were spread wide. She was agitated. Seventy yards might sound like a long distance, but in personal space for an elephant, it isn't. Trumpeting vehemently, she charged.

I turned and ran like a fool. Mwaniki turned and ran like a gazelle. Douglas-Hamilton turned and ran—then thought better of it, turned back, threw his arms out, and hollered to stop her. Sometimes this works; some elephants (such as old Boadicea, back at Manyara) make bluff charges, or halfhearted ones, and can be halted by a gutsy challenge. But this charge wasn't bluff. The female honked again and kept coming. Douglas-Hamilton turned again and ran.

By now I had a 20-step lead and Mwaniki was gone. At the rate he'd been moving, he might have been halfway to Lamu. But no: He ran straight back into camp (we learned afterward) and shouted in Samburu: *"Etara lpayian ltome!*—The old man has been killed by an elephant!" This announcement, though premature, brought people back to the scene fast.

Meanwhile the elephant caught Douglas-Hamilton as he tried to evade her by circling a bush. From 50 feet away, I watched her lift him with her trunk and then toss him, as you'd toss dirt off a shovel. He uttered a single piteous word, "Help," as he flew through the air. She stepped forward and stabbed her tusks downward. Douglas-Hamilton's body was now obscured by tall grass, and I couldn't see whether she had skewered him. Then she backed off about 10 steps and paused. This was the moment, he told me later, when he had time to wonder whether he would die.

She turned away, shuffling off a few dozen strides.

I ran to Douglas-Hamilton, feeling responsible, expecting gore, but to my relief, his innards weren't spilled out like ratatouille. Careful, David, she's still there, he said. But now the elephant's aggravation was evidently spent, and she marched away to find her calves. Iain sat up. He was scratched, dazed, bruised, rumpled. His shoes, glasses, and watch were gone, but he was intact. Beside him in the sandy clay soil were two tusk holes, rammed in about seven inches deep. I felt all over his rib cage: no tusk wounds. She couldn't have come closer to nailing him to the ground like a tent, but had she missed, or had she just wanted to make a point?

I helped him to his feet. And then a dozen people arrived, running and driving from camp. Someone found his glasses and shoes. The watch was busted but ticking. Quickly we vacated the area, before the elephant could change her mind and come back.

In the aftermath Douglas-Hamilton and I pieced together what had happened. We differed on the matter of her aim and intent. I thought she

THE HEARTBEAT OF THE WILD

had missed him. He was quite sure that she had not—not by mistake. No, if she'd wanted to kill him, he insisted, she would have. There was much anxious release, much apologizing (especially by me, for getting us out there on foot, but he wouldn't hear of that and claimed the blame himself), and much hypothesizing. With help from Daballen and Lentipo, he established that this female must have been Diana, of the Royals, with her two youngish calves. Maybe we startled her because the wind had been at her back; therefore she couldn't smell us before we got near. Maybe she feared for her calves. Maybe she had been put on edge by a pushy bull, or a lion, just before we blundered along. Is there anything in the records on Diana, he asked his people, that would suggest a recalcitrant disposition? There was not.

Diana. She was "just" another elephant: sensitive, volatile, and complex. Her behavior that afternoon, though violent, had been nuanced. At the last moment, it seems, she made a choice. She chose not to kill him, he believed. And no one, not even Iain Douglas-Hamilton, with all his magical gadgetry and his hard-won knowledge, will ever know why.

· ·

Years pass, decades pass, but elephants can live a long time—if they are fortunate enough to elude the perils of accidental death, lion predation, poachers seeking ivory, and the many other hazards of coexisting with humans. Research projects can last a long time also— and when it comes to research on the behavior of long-lived animals, continuity and duration are crucial. The work of Save the Elephants goes on, at Samburu National Reserve and elsewhere, 30 years after its founding and almost 60 years after Iain Douglas-Hamilton established his first study near Lake Manyara. He remains the senior scientist as well as the founder and guiding spirit of STE; his activity in recent years has been focused more than ever on mentoring the next generation of elephant field scientists and conservationists, and on speaking to audiences worldwide about the continuing crisis of elephant losses.

Some of the figures in my story have moved on, but David Daballen is now head of field operations for STE; Fritz Vollrath still chairs the board of trustees; and George Wittemyer, an associate professor now at Colorado State University, heads the STE scientific board. The female elephant known as Diana has long since died, but some of her offspring remain. Maybe that includes one of the two calves that witnessed her chastisement of Iain and me and Mwaniki.

Unfortunately, the devastating trend of population decline among African elephants continues too. Poaching for ivory and loss of habitat, all across elephant distributional range as lands are converted to agriculture and other uses, remain two of the main drivers of population decline. The savanna elephant (Loxodonta africana, the species of open habitats such as Samburu) is now classified as endangered on the Red List of the International Union for Conservation of Nature (IUCN). The forest elephant (Loxodonta cyclotis, of dense moist-forest habitats such as those in ROC and Gabon) is now listed as critically endangered. The rate of losses to poaching has declined somewhat, since a peak in the years 2008–2011, but total numbers are still falling too. An ambitious continent-wide estimate compiled under IUCN auspices in 2016, from multiple data sources, reported a loss of roughly 111,000 elephants during the preceding decade. The continental total was put at about 415,000 animals, counting both species. Africa had once, as recently as 1930, contained somewhere up to 10 million elephants.

But in 2017, a very positive development occurred: After years of international persuasion and a complex change of attitudes, China banned all ivory trade within the country. "The Chinese finally got it," Douglas-Hamilton told me recently. There were many contributing factors, economic and cultural, but it certainly didn't hurt that Yao Ming, the Chinese basketball star, had visited the Samburu National Reserve several years earlier and appeared in an antipoaching film. "In China, when Yao Ming speaks, people listen," according to an account in the New York Times.

The effect of the China ban on ivory commerce was vast, and felt even in ecosystems such as Samburu, where the presence of scientific study teams and an increasing appreciation of elephants among local people had already somewhat mitigated the poaching. Other threats to the Samburu elephants, less dramatic and abrupt than gangs of poachers with Kalashnikovs, remain serious as well—degradation of the habitat by an increasing abundance of livestock, for instance.

"This is a very dry and fragile land," Douglas-Hamilton reminded me. The Samburu by tradition are seminomadic pastoralists, herding cattle, goats, and other stock, which at some times and places compete with elephants for graze and browse. How to mitigate that competition? By showing respect, contributing toward fixes, and offering options. Even for a people as grounded as the Samburu, there can be other forms of wealth and dignity besides owning cows. It's the same conundrum that exists in many other landscapes, and Douglas-Hamilton and his team (largely consisting now of smart young Kenyans) are sensitive to it. They know well that conserving wildlife for the long term must also involve considering, and supporting, the livelihoods of local people among whom wild animals live.

THE LONG
WAY HOME

Salmon, Politics, and Sustenance on Russia's Kamchatka Peninsula

....

I made two trips to eastern Russia, in August of successive summers, to research this story about salmon and river conservation on the Kamchatka Peninsula. I saw a panorama of riverscapes and wild countryside and volcanoes, traveling to and over those places on the huge, rattling Mi-8 helicopters that are virtually the only form of aviation in Kamchatka. The food wasn't great, except for the fresh salmon, which was abundant, cheap, and delicious. My sense of guilt about eating so well, at the expense of these magnificent fish, was lessened by my knowledge that all Pacific salmon are semelparous, which means—as I explain in this piece—they have a longish life of growth, one act of reproduction, then death. Being caught on a river and eaten, by a bear or a human, stops them short of reproducing but only reduces the life span of an individual fish by weeks. And because each female is very fecund, producing so many eggs, so many fry, most of which perish before adulthood, a well-regulated fishery can remove a fraction of the spawning run without reducing population numbers. Spoiled by availability, I developed a fondness for salmon roe (loosely called caviar, although that term more strictly applies only to the salt-cured eggs, or roe, of sturgeon-family fishes). Whatever you call the stuff, that close to the source, it was exquisite.

Lodging for some days at a small hotel in a village outside Petropavlovsk, I also admired the volcanoes looming from the surrounding horizon, and I noted with perplexity the huge, stump-legged statue of Lenin, with flaring overcoat, that glowered onto a public square near the waterfront. Why was he still there—unregenerate Marxism-Leninism, or campy nostalgia? I spent enough time calling on bureaucrats and activists in the city that I became comfortable with the Petropavlovsk bus system. But the splendor of Kamchatka begins when you get just beyond those city limits.

. . . .

The Kamchatka Peninsula, rugged and remote, is a vast blade of land stabbing southwestward through cold seas from the mainland of northeastern Russia. Its coastline is scalloped like the edges of an obsidian dagger. Its highlands rise to cone-shaped volcanic peaks, snow-streaked in summer, and to ridges of bare, gray rock. Its gentler slopes are upholstered in boreal greens. It's a wild place, in which brown bears and Steller's sea eagles thrive on a diet rich in fatty fish. About 312,000 people currently inhabit Kamchatka Krai (its label as a governmental region), or less than a single person per square mile, and on a declining trend, but the humans too are highly dependent on fish. In fact, you can't begin to understand Kamchatka without considering one important genus: *Oncorhynchus,* encompassing the six species of Pacific salmon.

Then again, it might also be said: You can't understand the status and prospects of *Oncorhynchus* on Earth without considering Kamchatka, the secret outback where at least 20 percent of all wild Pacific salmon go to spawn.

Although larger than California, the peninsula has less than 200 miles of paved roads. The capital is Petropavlovsk-Kamchatskiy (known to locals as P-K and to the rest of us simply as Petropavlovsk), on the southeastern coast, containing half the total population. Across a nicely protective bay sits the Rybachiy Nuclear Submarine Base, Russia's largest, in support of which the city grew during Soviet times, when the entire peninsula was a closed military region. Travel to most other parts of Kamchatka is still difficult for anyone who doesn't have access to an Mi-8 helicopter. But there is a modest network of gravel roads, and one of those winds upstream along a narrow waterway called the Bystraya River, amid the southern

Central Range, to the Malki salmon hatchery, a complex of low buildings surrounded by trees.

Hatchery operations began in Kamchatka in 1914, during the twilight of the tsars, but this facility was established a life span later. In a lounge room off the entryway, by the time I got there, someone had hung a poster, declaring in Russian: "Kamchatka was created by nature as if for the very reproduction of salmon." That sounds almost like a myth of origins, but the poster listed some nonmythic contributing factors: Permafrost is largely absent, rain is abundant, drainage is good and steady, and because of Kamchatka's isolation from mainland river systems, its streams are relatively depauperate of other freshwater fish, leaving *Oncorhynchus* species to face few competitors and predators. The poster was right. Judged on physical and ecological grounds, it's salmon heaven.

Unfortunately, those aren't the only factors that apply. Kamchatka's tottering post-Soviet economy, fisheries-management decisions (and the politics behind them), and how those decisions are enforced will determine the fate of Kamchatka's salmon runs, driving them toward a future that lies somewhere between two extremes. Within a relatively short period, maybe 10 or 20 years, the phrase "Kamchatka salmon" could represent a byword for good resource governance and a green brand, reflecting the greatest success story in the history of fisheries management. Or that phrase could memorialize a squandered conservation opportunity of the early 21st century. Think: American alligator. Or think: passenger pigeon. At present, the situation is fluid.

Life is hard enough for a salmon, even without politics and economics. Consider the 1.2 million fry released each spring from the Malki hatchery. Roughly five inches long after their first months of growth, they face no easy path from infancy to adulthood. What they face, rather, is a high likelihood of early death. For starters, the hatchery lies about a hundred miles (as a fish swims) from the sea. Each little salmon must descend the Bystraya River to its confluence with a larger river, the Bolshaya. Eluding all manner of freshwater perils on the Bolshaya, it must gradually metamorphose into a different sort of fish, a smolt, capable of making the transition to life in salt water.

From the mouth of the Bolshaya, on Kamchatka's west coast, it must enter the bigger world of the Sea of Okhotsk, a frigid but nourishing body of water between the peninsula and mainland Russia.

Then, for a period of two to five years (depending on the species), that fish must circulate through the Sea of Okhotsk or else southeastward around the peninsula's tip into the expanse of the Pacific. The fish might travel thousands of miles, finding its preferred food (mostly small squid and crustaceans) abundant, but also facing predation, competition, and other challenges of the marine environment. For instance, it might be caught in the open ocean by fishermen using enormous drift nets that trap everything in their path. If it survives these years of robust swimming and feeding, it will grow large, fat, and strong. That's the advantage of anadromy (a sea-run life history): The ocean years allow fast growth. Approaching sexual maturity, the fish will head homeward to spawn, using some combination of magnetic sensing and polarized light to find its way back to the Bolshaya River. From the estuary it will ascend upstream by smell, branching into the familiar Bystraya, and finally climbing through the same shallow riffles of the same smaller tributary that its parents ascended before it.

Thousands of eggs will be laid for every two adult fish that return. Unlike an Atlantic salmon or most other species of vertebrate, a Pacific salmon breeds once and then dies. Scientists call the phenomenon semelparity. For the rest of us: big bang reproduction. After the adult has homed to its spawning stream, death follows sex as inexorably as digestion follows a meal. It's a life-history strategy, shaped by evolution over millions of years, that balances the costs of each spawning journey against the costs of reproductive effort, toward the goal of maximizing reproductive success. In plainer words: Because the likelihood of any fish surviving the whole journey not just once but twice is so slim, Pacific salmon exhaust themselves fatally—they breed themselves to death—at the first opportunity they get. Why hold back anything if you'll never have another chance?

So their lives enact a romantic but pitiless narrative. Their success rate is low, even under optimal circumstances. The miracle of salmon is that any of them manage to complete such an arduous cycle at all. And present circumstances on the Bolshaya River and its tributaries—though the wall poster at Malki didn't say so—are far from optimal.

LUDMILA SAKHAROVSKAYA, DIRECTOR of the Malki hatchery at the time of my visit, was a sweet-spirited woman with blond hair and silver glasses who had worked there since the early 1980s. She trained as a biologist in Irkutsk, a warmish city in south-central Siberia, before moving east to this severe outpost in search of a better livelihood. For almost three decades she watched, like a doting nanny, the cycles of salmon rearing, release, and return.

"Twenty years ago, I remember lots of fish coming to this river," she told me through a translator on a crisp summer day as we stood near her fish traps in a little tributary. Those traps were the end point for spawn-ready adults whose eggs and sperm would fuel the hatching and rearing operations of the hatchery. "A variety of species," Sakharovskaya said. "Now I don't see them."

The decline in the run of chinook *(Oncorhynchus tshawytscha)* has been especially severe, she said. These are deep-bodied and silvery creatures with purplish dorsal markings, largest of all salmon species, and therefore sometimes known as king salmon. Once they came in great, regal herds. Nowadays the Malki hatchery releases 850,000 chinook fry (as well as a lesser number of sockeye) annually, but not many adults return. What stops them? Two kinds of illegal harvest: overcatching *(perelov* is the Russian word) by licensed companies that have catch quotas but exceed them with impunity, and poaching by individuals or small crews, mostly for roe (salmon eggs, to be processed as caviar), at concealed spots along the river. The poaching problem throughout Kamchatka is catastrophic in scale, totaling at least 120 million pounds of salmon annually, much of it controlled by criminal syndicates. A hatchery director cannot fix that problem, Sakharovskaya noted, and the regulatory authorities evidently don't have the resources or the resolve to do it. So only the luckiest and most elusive of chinook reach their destiny here along the Bystraya. "We can almost count them on fingers," she said.

Does that make you sad? I asked.

"Da."

BUT THE BOLSHAYA drainage is only one of many river systems on the peninsula, and its hatchery fish aren't representative of Kamchatka wild salmon. Circumstances elsewhere are different; threats, opportunities, regulations, and even bureaucratic structures all change year by year. The whole situation is as complicated as a nested set of matryoshka dolls—Putin containing Gorbachev containing Brezhnev containing Stalin. On the Kol River, for instance, which also drains to the west coast, there is no hatchery, no streamside road, and (so far) no tragedy of depleted runs. What the Kol represents is superb habitat, scarcely touched, and abundant runs of wild salmon, including all six species: chinook, sockeye, chum, coho, pink, and masu. In 2008, more than seven million fish returned to spawn, filling the Kol so fully that, along some stretches, salmon were packed side to side like paving bricks. The Kol also carries another distinction. By a 2006 decree of the Kamchatka government, that river (along with another nearby stream) became part of the Kol-Kekhta Regional Experimental Salmon Reserve, the world's first whole-basin refuge established for the conservation of Pacific salmon.

On the north bank sat the Kol River Biostation, a cluster of simple wooden buildings that served as base for a binational research effort, its field operations led by Kirill Kuzishchin of Moscow State University and his American colleague, Jack Stanford of the University of Montana. Kuzishchin, Stanford, and their team were studying the dynamics of the Kol ecosystem. They hoped to address several big questions, including: How important are salmon to the health of the entire river ecosystem?

Kirill Kuzishchin, a burly man with a linebacker's neck and a sharp scientific brain, was raised on a farm near Moscow by his grandparents. At age four he caught his first fish and was evermore fascinated by things piscine; even in adulthood, as an associate professor in the ichthyology department at Moscow State, he loved to cast a line when collection of specimens was required. Among the chief lessons of his studies in freshwater ecology was that a river is more than its main channel. "The whole flood-plain acts as one single organism," Kuzishchin told me during a late evening talk at the Kol station. Water flows not just downstream but from channel to channel, both on the surface and via the underground aquifer; leaves fall into the river from riparian trees and bushes, supplying food and mineral nutrients to aquatic insects and microbes; whole trees topple into the water, providing cover for fish. "Everything is connected," he said. "The faster the growth of the trees, the more of them falling down into the river, the more habitats we have."

But nutrients are continually lost from the upper reaches by the same gravitational pull that takes water, silt, and other material downstream. So why don't these rivers gradually lose productivity? The reason is upstream migration by millions of salmon, Kuzishchin explained. The fish themselves bring nutrients such as nitrogen and phosphorus, accumulated during the years at sea, and surrender those precious loads to the ecosystem as their bodies decay. One aspect of the fieldwork by his and Stanford's team was to gauge the amount of nutrients at the Kol headwaters that salmon deliver and redeliver.

The scientists did that by testing for a certain isotope of nitrogen, N-15, which is relatively rare compared with other forms of nitrogen but far more abundant in oceans than in rivers. High concentrations of N-15 in the Kol's water during the season after spawning and decay, and in the leaves of the willow and cottonwood trees lining the banks, reflect the fact that salmon are bringing nutrients upstream. It's a circular effect, Kuzishchin said. Take the salmon away (for instance, by overharvesting or by poaching) and the very leaves of the trees will be deprived of nitrogen. So will the microbes and insects that eat the leaves. The entire ecosystem will lose nutrients, possibly to the point that it can no longer support large runs of salmon, even if they were reintroduced. He repeated his ecological maxim: "Everything is connected."

Jack Stanford made the same point more bluntly: "If you harvest all your fish, you cannot have a productive system."

During a fieldwork excursion upstream from the station, I saw the Kol River's fecundity for myself. We ascended the main channel in motorized johnboats, then proceeded by foot, bashing across the jungly floodplain through a dense thicket of green plants, 10 feet high but as delicate as celery, toward a side channel where the team would gather data. Kuzishchin led, chopping a corridor through the vegetation. It was mostly annual growth, consisting of thistle, nettle, cow parsnip, and a white-flowered thing called Kamchatka meadowsweet, along with some grasses and ferns, together constituting a fast-growing floodplain assemblage known in Kamchatka as *shelomainik*. Finally, we reached the little spring-fed channel, and as Kuzishchin and the others prepared to collect stream insects, algae, tiny fish, depth and flow readings, and willow leaves for nitrogen testing, I asked Stanford: What allows all this herby growth to spring up here, within such a short growing season, every year?

"In a word," he said, "salmon."

SALMON SUPPORT A human ecosystem too. On the lower Bolshaya River, three hours west of Petropavlovsk along another of those gravel roads, is the town of Ust-Bolsheretsk, once a robust fishing-and-processing entrepot. I reached it, with my driver and translator, after only a brief stop at the military checkpoint, where bored soldiers brushed away mosquitoes and didn't bother to inspect our van; we were headed toward the salmon, after all, not away from the salmon and toward the black market in Petropavlovsk.

The vice-mayor of Ust-Bolsheretsk District, Sergei Pasmurov, received us in a sparsely furnished office behind a double set of riveted, leather-padded doors and beneath a photo of Vladimir Putin glowering down from behind a fern. Pasmurov, a white-haired though youngish man with a bristly gray mustache, keen brown eyes, three gold incisors that flashed when he spoke, and ramrod-straight bearing, bore a strong resemblance to the novelist William Faulkner, if Faulkner had gone to a Soviet dentist. Pasmurov offered a candid sketch of local history, which had been difficult recently.

Throughout the Soviet era, Ust-Bolsheretsk was a sizable agricultural center, he explained, a base for several large collective farms that kept dairy cattle and raised turnips, tomatoes, and other vegetables in hothouses. Fishing was secondary, with two fish-processing plants operating here. Population stood at about 15,000 for the district, including Ukrainians, Belarusians, Irkutsk Siberians—people from all over the former Soviet

Union—as well as Kamchatkans of the Itelmen ethnic group. Then, so abruptly, so harshly, came the end of the U.S.S.R., without which those government-supported agricultural collectives failed. Suddenly there was an unfamiliar phenomenon, unemployment, and the population fell by a third. Dairy production dropped; vegetables became scarce. Pasmurov described the whole cascade of changes concisely and bundled them into one freighted word—I think it was *razvalilsya*—that my translator rendered as "the ruining." That is: disaster, collapse, ruination. Fishing became, for lack of alternatives, the major economic activity of the district.

But fishing was seasonal, also cyclical, with up-and-down fluctuations from year to year. And fishing was finite; even during a good year the river couldn't support everyone. Nonetheless, 18 different companies or individuals were licensed to fish thereabouts, the vice-mayor told me. "It results in reducing fish." Year to year, he said, the runs were becoming smaller. "Poaching also influences this process." The Bolshaya is a large river, easily accessible by road, therefore hard to protect. Access would become easier still, he added, now that a pipeline was being built to carry natural gas from the west coast to Petropavlovsk, crossing the headwaters of the Bolshaya and five or six other major rivers (including the Kol, notwithstanding its protected status). The pipeline itself might be clean and leak-proof, said Pasmurov, but the road built alongside it would invite more poaching, especially for roe.

And roe—preservable, compact, portable, very valuable as caviar—or *ikra,* as it's known in Russian, is what most poachers are after. "It's more convenient, easier to hide," he said. "You just salt it, put it in tanks, hide it in the forest." Later a truck, or even a helicopter, would come to collect the stash. Netting the salmon as they neared their spawning streams, slitting them open, stripping out the eggs, tossing the carcasses aside as waste, a gang of poachers could do huge damage in a very short time. Their wholesale customers might even include some of the big fish-processing plants: laundering caviar for the open market.

I had heard the same thing about caviar from other sources, including an ex-poacher on the Kol, who recalled that in the old days a small team could harvest five metric tons in a season; that there might be 15 such teams on the river; and that each man would make 10 times the money he could in a legal fishing job. Do the arithmetic and you found that 75 metric tons of illegally taken salmon eggs (each mass of eggs accounting for just 3 percent of the female's body weight) amounted to almost six million pounds of illegally killed fish. The carcasses of those salmon were left for bears and other scavengers—a short-term benefit to the ecosystem, yes—but every salmon thus intercepted left no offspring whatsoever to perpetuate the run.

PARTS OF THE human ecosystem too, in an inexorable process since the collapse of the Soviet Union, have experienced their own form of razvalilsya, the ruining. I saw that after our meeting with Mayor Pasmurov, when we drove west another 15 miles to the coast, on the chilly Sea of Okhotsk, then south along a narrow, fragile cape that was barely more than a long stem of sand between the Bolshaya estuary on our left and the sea on our right. We came there to a decrepit, half-abandoned town with the cheerily patriotic name Oktyabrsky, as in the October Revolution of 1917, the one that brought Lenin's little gang of Bolsheviks into power. This place, once also a busy fishing town, now looked like the bombed-out scene of an artillery battle. There were five-story apartment blocks in the concrete Soviet style, with units that seemed abandoned, and other buildings half-demolished or, where the waves undercut the sandy terrain, literally falling westward into the sea. Garbage blew off a bulldozed landfill, iron boats and car bodies lay gathering rust, dogs prowled loose amid the smell of rotting salmon. Farther south, we found two private fishery companies still operating, in steel buildings behind fences. A sign, roughly whitewashed onto the side of a shed, offered *"KUPLI KIZUCH I IKRU*—I will buy silver salmon and caviar." An arrow pointed toward a backstreet garage. Nearby was a concrete bust of Lenin, not such a bold effigy as the one that still stands over the city square in Petropavlovsk, glowering down at the tourists who take its photo as an anachronistic curio. This one sat neglected amid overgrowing grass.

The place intrigued me, so the next day we returned to Oktyabrsky for an interview with the manager of one of those fish-processing plants. We found him on the second floor. His name was Nikolai Ageev, and he bore the nose of an old prizefighter who had faced mostly left-handed opponents. He told me about the fishing season and the size of his workforce. When I asked about the town itself, he motioned toward a henna-haired woman seated quietly at the desk opposite: Elena Viktorovna Scherbal, who could speak of this woebegone town because she had the misfortune to be born here. (Henna, I had noticed, was quite fashionable as a hair color among workingwomen in eastern Russia at this period.) Scherbal wore a print dress, a sly smile, and a mole on the left side of her upper lip—I kid you not, almost like Miss Kitty in *Gunsmoke*. How has Oktyabrsky changed since the early 1990s? I asked her. In the worst way, she said. The population is half what it was, and most of those people leave when the fishing season ends. They keep apartments here, she said, but—and at that point our conversation was interrupted by noisy chatter from others and the clatter of the fish-processing line downstairs.

I was curious about that too, the processing operation, but unfortunately it was pollack, not salmon, being gutted and cleaned and packed. The night shift would do salmon, starting around 11:30 p.m., the manager volunteered. We could come back to see that if we wanted, he said.

As we left, that was the farthest thing from my mind. But over dinner back in Ust-Bolsheretsk, eating *shashlik,* I decided that I wanted to know more about Elena Viktorovna Scherbal and her town, Oktyabrsky. Get some rest, I told the translator and the driver, we're going out there again at 11:30.

By midnight we were back in the factory, meeting a sharp-featured young man named Denis, who ran the caviar room, and his boss, the imposing but affable Pyotr Viktorovitch Korenovsky, who stood about six foot seven and bore the title "Chief Master of Shift." In a hoodie sweatshirt, coveralls, and a buzz-cut hairdo, Korenovsky moved swiftly from one station to another along the production line and unloading area, supervising, lending a hand, chewing gum, and explaining it all to me (through the translator) as he went. In my mind, he became Peter the Tall, a man who loved his work, or at least loved making the night shift efficient, and I scribbled a crazed flurry of notes about how to factory-process salmon.

These were mainly sockeye, he told me, along with a few chum salmon and arctic char. He explained the proper procedures by which they were cut, gutted, cleaned, and sorted by size, species, and grade of quality. How they were weighed, frozen, and packed. And how, before any of that, they were caught, legally and illegally. Yes, there is a decline in the runs, he said. The problem was not unsustainable quotas, in his view. The problem was poaching. I learned that he was 29 years old, and still a fisheries student in the off-season. This was to be his life. I learned that he ate no salmon himself. I didn't learn why not.

Then, at 1:25 a.m., against any expectations or hopes, my translator spotted Elena Viktorovna Scherbal, just off the packing floor, and we caught her. Yes, she works all hours this time of year, she said. She agreed to a brief chat, back upstairs, away from the din. She told me a bit about life in Oktyabrsky, hers in particular—that her father had come to the town as a seasonal fisherman, during the Soviet era, and stayed as a fireman; that she began on the fish-processing line herself at age 18; that she now had a 14-year-old son, whom she and her husband hoped might enter customs work, not fisheries. She spoke of how she and others survived the hard times of the early 1990s, just after the U.S.S.R. went kaput, when the national government was just gone, missing in action, and there was no central heating, no jobs, no electricity. How did they get by? "Lots of candles," she said.

Before we departed, Peter the Tall reappeared and gave us a guided tour of the factory's sanctum sanctorum, the caviar room. He and Denis explained the fastidious steps of turning salmon eggs into delicacy: the straining, rinsing, salting, spinning, inspection, bucketing. One of the inspecting and bucketing ladies (henna-haired, again) offered me a handful of very fresh caviar. Salty, fatty orange globes the size of BBs, inexpressibly good. Then we said fervent *spasibas* to everyone, woke up the driver, and headed back to Ust-Bolsheretsk.

ON THE LOWER Bolshaya I saw the business of salmon fishing as practiced legally. It was a chilly July morning. A dozen men stood ready in waders, wool caps, and rubber gloves as a net was stretched far out across the river's channel by motorboat, then swung gradually downstream and drawn closed against the same bank, trapping hundreds of sockeye and coho in a watery corral. The men began walking the net back upstream, drawing it tighter toward the shallows, herding salmon onto a gravel beach. The fish, big and silvery (not yet flushed red, as they would be if they reached their spawning grounds), flopped robustly until there was nowhere to go. The men lifted them by their tails, one by one, and tossed them into a cargo boat. When that boat was full, it departed upstream to a landing, for unloading onto a truck, and another took its place. Within half an hour, from one set of the net, the crew took what looked like at least a thousand pounds of fish.

At one point a man lifted a nice-looking fish by the tail and tossed it back into the river. It was a female, heading upstream to spawn, I was told, so they didn't want to kill it. One fish, at least, might fulfill her reproductive potential. But whether this operation would abide by its legal quota was another issue, impossible for me to judge as a casual observer.

LONG BEFORE THE Russian residents of Kamchatka (who are immigrants and descendants of immigrants) came to depend on salmon fishing as a pillar of the economy, the Itelmen people and other Kamchatkans had developed cultural, religious, and subsistence practices centered on salmon. The Itelmen made their settlements along the banks of rivers, mostly in the southern two-thirds of the peninsula, where they harvested salmon using fish traps and weirs. They dried the pink flesh, smoked it, fermented fish heads in barrels. "Those fermented heads had a great deal of vitamins," one Itelmen elder told me. "They cleaned out your stomach and all the bad things in your body." The Itelmen even venerated a god, known as Khantai, whose iconic representation was half fish and half human. In autumn the people set up a tall, wooden Khantai idol facing the river, offered it oblations, then celebrated a harvest festival of thanksgiving for the fish that had come and supplication for more to follow.

A revived form of that ancient festival is held each year in a small settlement called Kovran, on the west coast, which now constitutes the capital of Itelmen life. Kovran villagers still fish by the traditional methods. But starting in the Soviet era, other things changed, threatening Itelmen traditions while bringing little relief to their hard lives.

Irena Kvasova, an Itelmen activist I met at a small office in a backstreet of Petropavlovsk, told me that Soviet policy had required country people, including her mother, to abandon their remote hamlets and aggregate in centers such as Kovran. There they became laborers on collective farms or fishing collectives, a very different existence for people accustomed to independence and living off the land. The Itelmen received tax exemptions, true, and the government bought the ferns and berries they gathered and the game they hunted, at fair prices.

But during the more fevered decades of Stalinism, Soviet authorities felt a need to find "enemies of the state." People anonymously denounced their neighbors, sometimes just to settle a grudge. Kvasova's own great-grandfather had been one victim. A proud Itelmen, a hunter and fisherman, leader of the collective council, he was targeted in a poison letter, arrested, and sent to the camps beyond Magadan—that is, the Kolyma River region, grimmest of all islands in the Gulag Archipelago—where he died. In the aftermath, her family stopped communicating, even with one another, to avoid attracting suspicion.

In the 1970s, as the Soviet regime became relatively less sinister but more stolidly bureaucratic, outsiders, arriving from the south, received positions and enjoyed benefits while the Itelmen people were marginalized in their own communities. Gorbachev's perestroika, followed by the collapse of the state, followed by orgiastic privatization, completed the process of dispossessing the Native people of lands, waters, and living resources they had husbanded for millennia.

One sign of that progressive dispossession is that the Itelmen must compete with commercial Russian fishing companies for limited salmon-fishing quotas. The Itelmen number only about 3,500 people, one percent of Kamchatka's total. Power resides in Petropavlovsk, not in Kovran, and power influences the granting of quotas and fishing sites from river to river. The bureaucrats who grant those quotas and sites on the Kovran River have been generous to outsider-owned commercial companies, an Itelmen leader named Oleg Zaporotsky told me, while restricting the local people to quotas that are marginal, even for subsistence. The companies employ a few Itelmen, but generally not in the better-paying jobs. And beyond the issue of subsistence fishing, Zaporotsky explained, some Itelmen want to establish their own fishing-and-processing cooperatives, which would bring income to the community, support schools and other institutions, and provide good jobs that would keep people from drifting away.

Zaporotsky himself had partnered with others to buy a diesel generator for freezing fish, the first step toward claiming a market share. "If we don't create some small enterprises," he told me, "these settlements won't survive." So far, the bureaucrats had refused to grant commercial-scale quotas to Zaporotsky's local group or any other.

ON THE OPPOSITE side of the Pacific, the wild salmon runs of North America (south of Alaska), once great, have been devastated—and in some cases obliterated—by dam building, dewatering for irrigation, overfishing, agricultural pollution, and other forms of habitat degradation, and they have been genetically diminished by reliance on hatcheries. The people of Kamchatka have a chance to be wise and provident where Americans and Canadians have been stupid and careless. For Kamchatka to become the world's foremost wild salmon refuge, the runs in its rivers don't need to be restored; they need only to be protected from poaching, overcatching, oil and gas spills, disruptive and poisonous mining, and other forms of shortsighted mistake. The region could also become one of the richest export producers of fresh salmon, frozen salmon, and caviar. Those two prospects aren't incompatible; they're interlocked.

This is why the Wild Fishes and Biodiversity Foundation (WFBF), of Kamchatka, and its American partner, the Wild Salmon Center (WSC), supplied help and encouragement when the Kamchatka government created the Kol-Kekhta Regional Experimental Salmon Reserve, and why they supported efforts toward designating another protected area for salmon, on the Utkholok River up north. Staff of the Wild Salmon Center, in

Portland, Oregon, and in particular its president, Guido Rahr, were very helpful to me before I went to Kamchatka. As I learned first from Rahr during a Portland stopover, WSC and WFBF also backed an ambitious vision of adding five more such protected areas—on the Oblukovina, Krutogorova, Kolpakova, and Opala Rivers (all draining to the west coast), and the Zhupanova River (draining east), each to encompass not just the river itself but also its full drainage basin, including the headwater streams in which the salmon spawn and all the terrestrial habitat. Those five areas, together with the Kol and the Utkholok, would make Kamchatka the planet's greatest, boldest experiment in nurturing wild salmon for their own sake and for the measured use of humankind. And it seemed possible—if long-term management perspectives informed by scientific research, along with honest governance backed by strict enforcement, were allowed to triumph over the scramble for short-term gain by insiders.

Of course, some people preferred the old way, whether that represented Soviet-flavored enterprises (the V. I. Lenin Fishing Collective still operates from a large building near the Petropavlovsk waterfront) or heedless private resource extraction in the spirit of Standard Oil, the Anaconda Company (copper), and Peabody Energy (coal). History and human need lay heavily on Kamchatka. Three decades after the dissolution of the U.S.S.R., that huge bronze statue of Lenin, thick-limbed and implacable, still stood in the plaza outside the main government offices. It must have seemed quaint to tourists (like the equally giant statue of Stalin I encountered once in Ulaanbaatar, Mongolia, where it loomed grumpily over the dance floor in a disco bar), but to others it still evoked the power of the central government. Moscow still set the course. People without jobs still needed to eat, and fish were there for the taking.

It's a long journey from idealistic plans to concrete, well-enforced protections, just as it's a long journey from the deep Pacific to the gravelly shallows of the Bystraya River. I couldn't forget what Ludmila Sakharovskaya said as we stood streamside at the Malki hatchery. Twenty-five years of hatching and nurturing fish, releasing them, and seeing ever fewer return had made her cynical. She was tired too, eager to take her pension and go to Irkutsk. Yes, we have reforms now, she said—or anyway, there's talk of reforms. But that's just talk, just formalities. Poaching is easy to do and hard to prevent. She knew of whole settlements, in the mountains, filled with people who could seek a legal job but didn't, who lived out the winter waiting for summer, when they could poach salmon.

Were things better, I asked, during the Soviet era?

She considered that for a second or two, and answered carefully: "Better for fish."

· ·

More than a dozen years later, the news about Kamchatka salmon is mixed. At the time I visited, in the summers of 2007 and 2008, an epidemic of poaching was underway—encompassing both the illegal harvest of fish, without license, and the unreported overcatching by private operators who had licenses and quotas. The five additional hoped-for protected areas for great rivers and their drainages—the Oblukovina, Krutogorova, Kolpakova, Opala, and Zhupanova—did not happen. The Kol River, which I visited with Jack Stanford, is still protected, harboring all six species of Pacific salmon plus three other Asian salmonids. "It's extraordinary," Guido Rahr told me. He is still president of the Wild Salmon Center in Portland, and in a recent conversation he lamented that surge in poaching, which increased to the point that half of all harvested salmon were being taken illegally. But he also described new initiatives and strategies that have helped reverse the trend.

"What we didn't get is parks," Rahr told me, referring to the protected areas that had been proposed for those other five rivers. Didn't get them, apart from the Kol, he meant. "But what we did get is a way to fight the poaching epidemic." Central to that fight is an independent nonprofit organization called the Marine Stewardship Council (MSC), launched in 1997 and now global in reach. The MSC operates a certification program to advise seafood buyers about which suppliers

operate legally and sustainably, abiding by quotas and participating in fishery improvement programs. It's the power of information in a market for precious goods. These precious goods—wild salmon and their eggs—are in demand from wholesale and retail customers increasingly aware that fisheries resources are finite, and that overfishing, either by bandits or by licensed but unscrupulous fishing companies who harvest more than their allotted quotas, can destroy not just salmon runs (bad enough) but also, as Jack Stanford explained to me, the ecosystems that they nourish.

The process of certification involves close independent scrutiny. The first of Kamchatka's salmon-fishing operations to clear that bar, in 2018, operates on the Kamchatka River and in Olyutorsky Bay, along the peninsula's far northeastern coast. By this point, 80 percent of Kamchatka's salmon fisheries are encompassed within that program. Critics note that, because MSC receives fees for the certification process over the industry it reviews, there exists a conflict of interest, and that in some cases MSC has certified unworthy operators on fisheries around the world. But even with flaws, MSC certification seems the best hope for preserving robust runs of wild Kamchatka salmon into the future.

Meanwhile, the impacts of large-scale mining also threaten the rivers and their fish. Beneath the Kamchatka landscape, from which rise almost 30 active volcanoes and hundreds more that have gone dormant, lies an abundance of valuable minerals and fossil-fuel resources: gold, silver, platinum, nickel, cobalt, natural gas, coal, and even peat worth harvesting. Ripping and pumping those treasures out of the ground and off the surface will involve great dangers to water quality. "Mining can kill a river," Guido Rahr said.

This story, like so many others involving the conservation of wild places and creatures, is a tale of suspense. The ending is not yet written.

LET IT BE

·╬·╬·╬·

The Splendid Isolation of
Kronotsky Nature Reserve

· ·

Before my first research journey to Kamchatka for the salmon story, my editor, Ollie Payne, asked me: Would you be willing to make a small side trip while you're over there? I asked where. To a remote nature reserve on the east coast, north of Petropavlovsk. Reason? A photographer named Michael Melford had been there and delivered to National Geographic a portfolio of gorgeous landscape images. Editorial wanted a short essay to accompany them. Sure, I said. What's the story angle? You'll think of something, Ollie said. When I got there, I found an extraordinary little outback full of lead-footed tourists like myself.

· ·

Some places on this planet are so wondrous, and so fragile, that maybe we just shouldn't go there. Maybe we should leave them alone and appreciate them from afar. Send a delegated observer who will absorb much, walk lightly, and report back, as Neil Armstrong did from the moon—and let the rest of us stay home. That paradox applies to Kronotsky Zapovednik, a remote nature reserve on the east side of Russia's Kamchatka Peninsula, along the Pacific coast a thousand miles north of Japan. It's a splendorous landscape, dynamic and rich, tumultuous and delicate, encompassing 2.8 million acres of volcanic mountains and forest and tundra and river bottoms as well as more than 700 brown bears, thickets of Siberian dwarf pine (with edible nuts for the bears) and relict Sakhalin fir trees *(Abies sachalinensis)* left in the wake of Pleistocene glaciers, a major rookery of Steller sea lions on the coast, a population of kokanee salmon in Kronotskoye Lake, along with sea-run salmon and steelhead in the rivers, eagles and gyrfalcons and wolverines and many other creatures—terrain altogether too good to be a mere destination. With so much to offer, so much at stake, so much that can be quickly damaged but (because of the high latitudes, the slow growth of plants, the intricacies of its geothermal underpinnings, the specialness of its ecosystems, the unsteadiness of its topographic repose) not quickly repaired, does Kronotsky need people, even as visitors? I raise this question, acutely aware that it may sound hypocritical, or anyway inconsistent, given that I've left my own boot prints in Kronotsky's yielding crust.

The government of Russia recognizes such spectacular fragility with that categorical status *zapovednik*, connoting roughly: *a restricted zone, set aside for the study and protection of flora and fauna and geology; tourism limited or forbidden; thanks for your interest, but go away.* It's a farsighted sort of statutory designation, bravely and judiciously antidemocratic in a country where antidemocracy has a long, brutal history. Scientists are permitted to enter zapovedniks, though only for research and under stringent conditions. Kronotsky is one of 101 such reserves in Russia, by the latest count, and was among the first, decreed in 1934. Before that it had been a refuge for sable (*Martes zibellina*, that Asian counterpart of the pine marten, with some of the world's silkiest, most sought-after fur), established in 1882 at the prompting of local people, hunters and trappers who valued the forests surrounding Kronotskoye Lake as prime sable habitat. The Kamchatka Peninsula is very distant from Moscow (as distant, in fact, as Moscow is from Boston), and to Joseph Stalin's Soviet government in the mid-1930s (with much else on its agenda), the opportunity costs of putting a modest chunk of that wilderness within protective boundaries probably didn't seem high. In 1941, a second kind of asset revealed itself within the reserve, when a hydrologist named Tatiana I. Ustinova discovered geysers there.

Ustinova, a doughty 27-year-old who previously worked in the Urals, had come to Kronotsky just a year earlier, along with her husband, a zoologist, who had been appointed acting director of the reserve. In the cold springtime of April 1941, Ustinova and her guide were exploring the headwaters of the Shumnaya River by dogsled. They paused near a confluence point and happened to notice, at some distance along the water's edge, a large outburst of steam. With hungry dogs and other urgencies pulling her away, Ustinova couldn't see much more, not then, but she returned in July to map and study what proved to be a whole complex of geothermal features, including about 40 geysers. She named her first geyser Pervenets, meaning "Firstborn." The tributary she ascended is now called the Geysernaya River, and above one of its bends is a slope known as Vitrazh (translated as Stained Glass), for its multicolored residue from a score of large and small vents. Kronotsky's Dolina Geyserov (Valley of Geysers) took its place as one of the world's major geyser areas, in a league with Yellowstone, El Tatio in Chile, Waiotapu on New Zealand's North Island, and Iceland.

Geysers are generally associated with volcanic activity, and that's certainly the case in Kronotsky. The entire Kamchatka Peninsula is abundantly pustulated with volcanoes, of which about two dozen, including some inactive ones, lie within this zapovednik or along its borders. Kronotsky Volcano is the tallest, a perfect cone rising to 11,552 feet. Krasheninnikov Volcano (named for Stepan Petrovich Krasheninnikov, a hardy naturalist who explored Kamchatka in the early 18th century) is its nonidentical twin, lying just southwestward across the Kronotskaya River. Still farther southwest is what would be, but no longer is, the third in a huge three-peak sequence. Instead of a high cone, this one is a broad, low bowl, up to eight miles in diameter, filled with fumaroles and hot springs and sulfurous lakes, blueberry-and-heather tundra, forest patches of birch and Siberian dwarf pine, all rimmed by a circular ridge left behind when a vast volcano blasted itself open about 40,000 years ago. The bowl is called Uzon Caldera. Its name comes from the kindly spirit Uzon, a powerful figure in the legends of the Koryak people. The exploration and study of Uzon Caldera by scientists, as well as Ustinova's finding of the Valley of Geysers, gave additional purpose to the zapovednik: protecting geological wonders as well as biological ones.

The story told by Koryaks about Uzon and his caldera has the ring of a parable. He was a friend to humanity, quieting Earth tremors, stifling volcanic eruptions with his hands, doing other good deeds. But he endured a lonely existence, living secretively atop his own mountain so that evil spirits wouldn't come and destroy the place. Then he fell in love. She was a human—a beautiful girl named Nayun, with eyes like stars, lips like cranberries, eyebrows as dark and glossy as two sables. She loved Uzon in return, and he took her away to his mountain. So far, so good. But after some years of marital bliss and isolation, Nayun began to yearn for her human relatives. Couldn't she have a visit with them somehow? Uzon, wanting to please her, made a desperate and tragic mistake: He spread the mountains with his mighty arms and created a road. People came, curious and disruptive. Now everybody knew Uzon's secret hideaway, including those evil spirits. "The earth yawned with a horrible crash having absorbed a huge mountain," in one telling of this tale, by G. A. Karpov, "and mighty Uzon turned into stone forever." You can see him there even today, petrified into a high peak on the northwestern perimeter of the caldera, his head bowed, his arms stretched around to form the rim.

If you do see him, you'll be among the few. The ban against tourism has been relaxed, but not much, for Kronotsky. About 3,000 nonscientific visitors enter each year, and of those, only half make a stop in Uzon Caldera. Regulations limit the number, but so do logistics, lack of infrastructure, and cost. For starters, there is no road into Kronotsky Zapovednik from the more settled parts (which are not very settled) of Kamchatka. No roads within the reserve either, notwithstanding the legend of Uzon. The in-and-out transport consists mainly of Mi-8 helicopters, thunderously powerful machines such as once ferried troops for the Soviet Army. Sitting in an Mi-8 as it powers up for takeoff, strapped into a rickety seat beside a port-hole window, you feel as you would in a crowded school bus with a sizable sawmill bolted to its roof—until the whole thing levitates. Tourist flights leave from a heliport 20 miles from Petropavlovsk, Kamchatka's capital, and are permitted to land only on helipads in the caldera and the Valley of Geysers. Neither place offers overnight accommodation for tourists, so a visit to the reserve constituted, at the time I went, a $700 day trip with lunch. The customer traffic seemed mostly made up of wealthy Russians,

Europeans on adventuresome holidays, and the occasional American. Five hours in Kronotsky wasn't something that ordinary families in Petropavlovsk could normally afford; it wasn't like loading the kids into the van for a summer trip that includes an ice cream at Old Faithful. Choppering in to see geysers and volcanoes and maybe a few brown bears (on my trip, the bears fled across the tundra as our pilot hazed them at low elevation, inexcusably, to provide us passengers a good look) was nature appreciation for the affluent, sedentary elite. It was dramatic and thrilling and privileged and rude. It made me crabby and dyspeptic to contemplate the extravagant wrongness of the experience, but . . . how would I know that if I hadn't been there and done it myself?

The authorities who managed Kronotsky and the scientists who studied it were sensitive to the downside of such tourism. Everybody left a footprint of some sort, the crucial questions being how deep and how many. At the beginning and the end of each summer season, investigators looked for impacts at the caldera and the geysers. Their reports helped inform decisions about the next season's visitation limits and dates. But the greater conundrum of Kronotsky, the one that provoked thought and not just sour belly, was how the concern over human-caused degradation should be reconciled with the inherent, violent dynamism of the place. This conundrum came to a point on June 3, 2007, when a massive wall of rock, mud, clay, and sand broke loose from a high ridge and slid, roaring, down a small creek valley, obliterating a hundred-foot waterfall, damming the Geysernaya River (in a matter of seconds), and burying much of the Valley of Geysers beneath the resulting new impoundment, which became known as Geysernoye Lake. George Patton's Third Army, marching through in hobnailed boots, couldn't have made such a mess. I arrived two months later, and saw the raw slopes above the Geysernaya River, the new lake, and the debris damming it, like a fresh gunshot wound that hadn't been cleaned or closed.

Pervenets, Ustinova's firstborn geyser, was gone. So were a few other famous spouts. The rest remained. Vitrazh, the stained-glass mosaic, was intact. Alarming reports had reached the international press, vacations had been canceled, and knowledgeable people were in disagreement about whether the slide was a tragedy or simply a fascinating natural shrug. "We scientists believe we are quite lucky to witness such an event," according to Alexander Petrovich Nikanorov, a researcher who briefed me at the zapovednik headquarters near Petropavlovsk. "Our lives are very short, and yet we witnessed it."

Geologists have good reason to feel that their lives, relative to the phenomena they study, are short. Rock usually moves slowly through time. But of course it's true for the rest of us also: Life is short, the world is big, and we're lucky to witness as much as we can. Whether that means we should all climb aboard the helicopter is another question, which I couldn't answer after the costly privilege of seeing Kronotsky—not even to my own satisfaction. What I could say (and what those photos by Michael Melford, as published in *National Geographic*, showed) was this: Kronotsky Zapovednik is an impressive place, fragile and magnificent and changeable. Maybe you can take that on faith?

....

Years after this grumpy meditation on Kronotsky and its beautiful, frangible landscape, I remained curious about the scientist who had discovered it and revealed it to the wider world, young Tatiana Ivanovna Ustinova. There's very little information on her available in English, and evidently not much in Russian either, but just enough to bring her story forward through time and then back—by a roundabout route—to the Valley of Geysers.

Ustinova and her guide, a local man named Anisifor Krupenin, almost died on that first expedition, according to one source. By the end of the first day, traveling by dogsled over the vegetation (still snow-crushed, even in April), without benefit of a trail, they reached a junction along the Shumnaya River where a sizable tributary, draining from one of the nearby volcanoes, entered. They made camp at that spot, but did not, as I had been led to understand when this piece was first published, happen to notice "at some distance" a venting of steam. What they did was decide to explore upstream along the tributary, and they set off next morning on skis, leaving their dogs, sleds, and other gear behind. The skis served poorly (presumably in wet, heavy snow on broken riverbank terrain), so they ditched them and proceeded in their high rubber boots, stumbling and falling as they went. The tributary wound on and on; drawing them forward now, besides curiosity, was a high column of steam rising somewhere ahead, possibly from a hot spring. Meanwhile it began to snow and blow, a spring blizzard. "The weather got worse and worse," Ustinova recalled, in her book Geysers of Kamchatka, published a decade later. They found no hot spring. They couldn't climb out of the valley. They decided to turn back—but first sat on the snow for a rest and a snack.

"Suddenly, from the opposite shore, a stream of boiling hot water shot straight at us from a small soaring platform, accompanied by puffs of steam and underground rumbling." It was scary at first—even, and maybe especially, for a geologist. "The behavior of volcanoes is unpredictable." Then the boiling eruption stopped, the steam cleared, the place fell silent. "That's when I came to my senses and shouted in a voice not my own, 'Geyser!!!'" They continued watching, and marked its regularity: two more eruptions, each lasting three minutes, at 45-minute intervals. This was the geyser she named Pervenets—"Firstborn."

She and Krupenin survived the night out there in a hand-dug snow cave. Next day they wandered. Krupenin was a seasoned guide, but this was unknown and confusing territory. On the third day—"frozen and barely alive"—they found a tent, evidently abandoned. That saved them, and they eventually wandered their way out. Ustinova, because she was tough as well as curious, but not stupid, returned—in July. She documented the geysers in detail. They are her monument, in more ways than one.

Back at the zapovednik base, Ustinova informed her husband about these wondrous discoveries, and he informed Moscow by telegraph. Moscow wasn't interested. Suspend all work, Moscow said, *and cut your expenditures to the bone. It was 1941, remember. Things had changed: On the morning of June 22, Hitler's army had invaded the Soviet Union and the war with Germany had begun.*

Ustinova and her family (including their first daughter, born during the war) remained in the Russian Far East until 1947, then moved to Crimea, because the daughter was ill, and then Moldova. Ustinova taught geology for some years at a polytechnic institute in Chișinău, the Moldovan capital. She returned to Kamchatka several times, for scientific research and other purposes. She wrote her book. After her husband died in 1987, she relocated to Canada, joining the elder daughter. She lived until age 96, and asked that her ashes be buried in Kronotsky, at the site of her geyser discovery. That happened in 2010, three years after the big landslide of June 3, 2007, which obliterated much of the Valley of Geysers as she had known it. Still, no tragedy: A geologist, better than anyone, would understand natural cataclysms of landscape as routine events in the geophysical metabolism of planet Earth. They put her remains in there somewhere.

Then, on January 4, 2014, another big landslide occurred, this one a collapse of the rim of another volcanic cone at the headwaters of the Geysernaya River. Dirt and rock and mud flowed. The water level of Geysernoye Lake shifted again, but not enough to obscure or radically alter the most elegant and famed of the geysers. Vitrazh, the slope punctuated by multiple small and large vents with their colorful residues, evocative of a stained-glass mosaic, is still there. The ashes of Tatiana Ivanovna Ustinova are still there too, though possibly not in the same circumstances where they were respectfully interred in 2010. As she would have understood, nothing on Earth is a permanent condition except change.

THE LEFT BANK APE

·|·|·|·

What Makes Bonobos Different

When you look at a map of Africa, if it shows geographical terrain as well as political boundaries, you will see three vast shapes that dominate the landscape—and have gone far to determine the history—of that continent. There's the Sahara Desert, spanning the northern third of the land mass and separating the Mediterranean world from Central Africa. There's the Great Rift Valley, a series of tectonic cracks and trenches running roughly north-south down the continent's eastern side, some stretches widening into long lakes, including Lake Turkana, Lake Tanganyika, and Lake Malawi. Finally, there's the Congo River, a vast avenue of water tracing a fishhook shape through the continent's midsection, from the Katanga Plateau to the Atlantic Ocean.

This river, which swells in size as it drains the equatorial forests on each side, is one of the most imposing geographical features on the planet. And because it curves through an arc of roughly 180 degrees—flowing north and then northeast and then west and then south and then west again—to speak of the "north bank" or the "south bank" can be ambiguous. There is a south bank of the river at Kisangani but only east and west banks at Lulonga. For clarity, it's better to think, as you face downstream, of the right bank and the left bank.

The river is big and forceful enough to divide biogeographical realms, not just countries. It flows through evolution, not just through landscape and history.

The place it defines, the Congo Basin, is challenging and irresistible. I have made it a policy that, whenever an editor asks, "Will you go to the Congo?" the answer is yes.

In a remote forest sector of the Democratic Republic of the Congo (DRC), along the Luo River, 50 miles by dirt trail from the nearest grass airstrip and roughly three times that far west of Kisangani, lies the Wamba research camp, a place quietly renowned in the annals of primatology. This field site (taking its name from the nearby village of Wamba) was founded in 1974 by a Japanese primatologist named Takayoshi Kano for the study of the bonobo *(Pan paniscus)*, a species of simian unlike any other.

The bonobo carries a reputation as the "make love, not war" member of the ape lineage, because it's thought to be far lustier and less bellicose than its close cousin, the chimpanzee. Modern studies of zoo populations by the Dutch American biologist Frans de Waal and others have documented its easy, pervasive sexuality and its propensity for amicable bonding (especially among females), in contrast with chimpanzee dominance battles (especially among males) and intergroup warfare. But the bonobo's behavior in the wild has been harder to know, and Kano, operating out of the Primate Research Institute of Kyoto University, was among the first scientists aspiring to study it amid its natural habitat in the Congo. Apart from several interruptions, including a hiatus during the Congolese wars of 1996–2002, observations at Wamba have continued ever since.

Early one morning I followed a researcher named Tetsuya Sakamaki, also from Kyoto University, into the forest at Wamba. Promptly I saw things that, according to the popular image of the species, I might not have expected. Bonobos quarreled. They hunted for meat. They went hours at a stretch without having sex. This was the animal so renowned for its lubricious, pacific social life?

As Sakamaki and I watched a party of bonobos feeding on the fruits of a boleka tree—small, grapelike morsels with papery husks—he identified the individuals by name. That female there, with the sexual swelling, we call her Nova, he said. She last gave birth four years ago; the gaudy inflation of her genital area, like a pink sofa cushion taped to her rump, advertised her readiness to breed again. This female is Nao, he said, very old, very senior. Nao has two daughters, of which the elder has so far remained in this group. And that female there, that's Kiku, also very senior, with three sons in the group. One of those sons is Nobita—easy to identify, Sakamaki explained, by his great size, by the absence of some digits from his right hand and both feet, and by the blackness of his testes. Missing digits suggested a mishap in a snare, not unusual for bonobos facing the hazards of human proximity. Nobita seemed to be the alpha male, insofar as bonobo groups recognize alpha males.

By now we had followed the bonobos into a grove of *Musanga* trees, and they were stuffing their mouths with a different fruit, pulpy and green. Suddenly a screechy altercation broke out between Nobita and another male, known as Jiro. Kiku, Nobita's elderly mother, charged over to support her son. Cowed by the two of them, Jiro retreated. He sulked in a nearby tree. It's interesting, Sakamaki noted, that Nobita is the largest male in this group, and yet his mother helps him in a fight. Even a high-ranking adult male such as Nobita seems to hold his status partly on the merits of his mama.

Forty minutes later, when the screeching began again, Sakamaki drew my attention to the focus of excitement: an anomalure (a gliding rodent, like a flying squirrel), scrambling for its life on a tree trunk while several bonobos converged around it. As the bonobos came closer, the anomalure launched itself into space and glided away. Then we noticed a second one, clinging secretively to the east side of another large bole while a bonobo named Jeudi sat clueless just 15 feet to the west. This anomalure, pink-eared and pale-eyed, held its place on the bark more patiently, frozen, not giving itself away. Within a moment, though, other bonobos spotted it, and the group closed in, shrieking with predatory menace. One bonobo climbed upward, struggling to find grips. The anomalure skittered 20 feet higher, ascending as easily as a gecko on a wall. When it was surrounded with bloodthirsty apes, the little rodent launched itself and sailed away through the limbs and undergrowth to safety. We never even saw where it hit the ground; neither did the bonobos. Wow, I thought. Nicely done.

"Hunting behavior—it's very rare," said Sakamaki. "So you are very lucky."

I did feel lucky. Not yet noon on my first day at Wamba, and already my notion of bonobos had been enriched and confounded with realities, contrasts, and complications.

BONOBOS HAVE BEEN confounding people ever since they first came to scientific attention. Back in 1927, a Belgian zoologist named Henri Schouteden examined the skull and skin of a peculiar animal, supposedly an adult female chimpanzee, from the Belgian Congo. The skull, he reported, was *"curieusement petit pour une bête de semblables dimensions*—oddly small for an animal of such size." The following year a German zoologist, Ernst Schwarz, visited Schouteden's museum and measured that skull as well as two others, concluding that they must represent a distinct form of chimp, unique to the south side—that is, the left bank—of the Congo River. Schwarz announced his discovery in a paper titled "Le Chimpanzé de la Rive Gauche du Congo." So from the beginning there was at least a subliminal association between the left bank culture at the center of the francophone world—the bohemian artists and writers and philosophers of *la rive gauche* in Paris, south of the Seine—and this newly identified, unconventional Congolese ape. Soon after, the left bank ape was recognized as a full species and took its modern scientific name, *Pan paniscus*.

Another label that fell upon the species was "pygmy chimpanzee," although it's almost as large as the common chimpanzee, the one already widely known *(Pan troglodytes)*. The bonobo's head is smaller in proportion to its body than a chimp's, its physique more slender, its legs longer. But in overall size, both male and female adult bonobos fall generally within the same weight range as female chimps. Scientists today tend to avoid the term "pygmy chimpanzee"; "bonobo" better suggests that this creature is not a miniaturized version of anything else.

The major distinctions between bonobos and chimps are behavioral, and the most conspicuous do involve sex. Both in captivity and in the wild, bonobos practice a remarkable diversity of sexual interactions. According to Frans de Waal: "Whereas the chimpanzee shows little variation in the sexual act, bonobos behave as if they have read the Kama Sutra, performing every position and variation one can imagine." For instance, they mate in the missionary position, something virtually unknown among chimpanzees. But their sexiness isn't just about mating. Most of those variations are sociosexual, meaning that they don't entail copulation between an adult male and an adult female during her fertile period. The range of partners includes adults of the same sex, an adult with a juvenile of either sex, or two juveniles together. The range of activities includes mouth-to-mouth kissing, oral sex, genital caressing by hand, penis-fencing by two males, male-on-male mounting, and genito-genital rubbing ("G-G rubbing" is the shorthand term) by two estrous females, who moosh their swollen vulvas back and forth against each other in a spate of feverish sisterly cordiality. Usually there's no orgasm culminating these activities. Their social purpose seems to be communication of various sorts: expression of goodwill, calming of excitement, greeting, tension relief, bonding, solicitation of food sharing, and reconciliation. To that list of benefits, we might also add sheer pleasure and (for the juveniles) instructional play. Varied and frequent and often nonchalant, sex is a widely applied social lubricant that helps keep bonobo politics amiable. De Waal again: "The chimpanzee resolves sexual issues with power; the bonobo resolves power issues with sex."

Sexiness isn't the only big difference between bonobos and chimps, though it's probably linked to other differences, either as cause or as effect. Females, not males, hold the highest social rankings, which they seem to achieve by affable social networking (such as G-G rubbing) rather than by forming temporary alliances and fighting, as male chimpanzees do. Bonobo communities don't wage violent wars against other bonobo communities adjacent to their territory. They forage during daytime in more stable and often larger parties, with sometimes as many as 15 or 20 individuals moving together from one source of food to another, and they cluster their nests at night, presumably for mutual security. Their diet, which is like the usual chimpanzee diet in most respects—fruit, leaves, a bit of animal protein when they can get it—differs in one signal way: Bonobos eat a lot of the herby vegetation that is abundant in all seasons—big reedy stuff resembling cornstalks, starchy tubers like arrowroot, nutritious shoots and young leaves and pith inside the stems, rich in protein and sugars. Bonobos, then, have an almost inexhaustible supply of reliable munchies. They don't experience lean times, hunger, and competition for food as acutely as chimpanzees do. That fact may have had important evolutionary implications.

Bonobos do share one distinction with chimpanzees: Together they are the two closest living relatives of humans. Back about seven million years ago, somewhere in the forests of equatorial Africa, lived a kind of proto-ape that was both their direct ancestor and ours. Then our lineage diverged from theirs, and by about 900,000 years ago, those two apes had diverged from each other. No one knows whether their last shared ancestor resembled a chimp, in anatomy and behavior, or a bonobo—but solving that uncertainty might say something about human origins too. Do we come from a long line of peace-loving, sex-happy, and female-dominant apes, or from a natural heritage of warfare, infanticide, and male dominance?

Also, what happened in evolutionary history to make the bonobo the unique creature it is?

RICHARD WRANGHAM PROPOSED a hypothesis. Wrangham, a distinguished biological anthropologist and a professor in the Department of Human Evolutionary Biology at Harvard, with more than four decades of experience studying primates in the wild, began his work on chimpanzees as a Ph.D. project, in the early 1970s, in Tanzania's Gombe Stream National Park, where Jane Goodall had pioneered the study of chimps. Wrangham continued at Kibale National Park in Uganda. No bonobos at either site; chimps and bonobos do not share habitat. But he addressed the subject of bonobo origins in a 1993 journal paper and then in a popular 1996 book, *Demonic Males*, co-authored with Dale Peterson. What led these two closely related apes, these two cousins within the genus *Pan*, to diverge into distinct species, rather than remaining just two populations of the same species in different places? The crucial point in Wrangham's hypothesis is the absence of gorillas, over the past one or two million years, from the left bank of the Congo River.

The reasons for that absence are uncertain, but the evolutionary consequences seem rather clear. On the river's right bank, where chimps and gorillas shared the forest, the gorillas ate what gorillas still eat, mainly herby vegetation, and the chimps ate a chimp diet, mainly fruit, tree leaves, insects, and occasionally meat. On the left bank dwelled that other chimp-ish animal, privileged by circumstance to be free of gorilla competition.

"And that's the formula," Wrangham told me by phone from his office at Harvard, "that makes a bonobo." The left bank creatures, bolstering themselves on a rich chimpanzee diet when available and sustained by those staple gorilla foods when it wasn't, lived a steadier life; they weren't forced to break into small and unstable foraging groups, diverging, rejoining, scrambling for precious but patchily available foods, as right bank chimps often are. And that fateful difference in food-finding strategy carried consequences for social behavior, Wrangham explained. The relative stability of foraging groups within a larger bonobo community meant that vulnerable individuals usually had allies present at any given moment. This tended to dampen dominance battles and fighting. "Specifically," he added, "females have other females as well as males available to protect them from those that might want to bully them."

Another result of the foraging-group stability, he noted, involves the sexual rhythms of bonobo females. Unlike chimp females, they aren't obliged by circumstance to present themselves always as extremely attractive, extremely ready for mating with all possible males during just short, periodic windows of time. "If you are a bonobo," Wrangham said, and you live in a larger and more stable foraging group, "then you can afford to have a long period of sexual swelling." A bonobo female doesn't need to attract gaggles of frantically horny males on a short-term basis. She's continually attractive, continually ready. "That greatly reduces the importance to the males of competing for dominance and bullying the females." So the famed amity and sexiness of bonobo social life had, by Wrangham's hypothesis, an unexpected source: the availability of gorilla foods uneaten by gorillas.

And why are gorillas absent on the left bank? Wrangham suggested a scenario, speculative, he acknowledged, but plausible. Sometime after about 2.5 million years ago, severe drying seems to have hit Central Africa. In the equatorial lowlands on both sides of the Congo River, herbaceous vegetation—gorilla habitat—shriveled away. Chimps could survive by finding fruit in riverside forests, but the right bank gorillas were forced into highland refuges, such as the Virunga volcanoes in the northeastern part of the drainage (where they evolved into mountain gorillas, a subspecies of the eastern gorilla) and the Cristal Mountains in the west. On the left bank, though, there were no such highland refuges. The land is flat. So if gorillas had ever lived on that side, the Pleistocene drought may have killed them off.

BONOBO BEHAVIOR IS exceptional among apes, and there are exceptions to the exceptions. You cannot paint their portrait with a broad brush. No researchers have been more punctilious about this than Gottfried Hohmann and Barbara Fruth, a married couple based at the Max Planck Institute for Evolutionary Anthropology, in Leipzig, Germany. Hohmann and Fruth have studied bonobos in the wild for more than two decades. Their work began in 1990 at a site called Lomako, in northern DRC, and they enjoyed uninterrupted field seasons until war started in 1998 and kiboshed everything for four years. The couple then established a new field camp farther south, at a place called Lui Kotale, in an excellent piece of forest just outside Salonga National Park. They arranged a compact with the local community within whose traditional territory the forest lies: In exchange for a fee, the local people agreed not to hunt or cut trees at Lui Kotale.

To get there, you land at another grass airstrip, walk an hour into a village, pay your respects to the elders, and then keep walking for five more hours. You cross the Lokoro River in a dugout canoe, wade up a black-water stream, climb a bank, and find yourself in a neat, simple camp of thatched ramadas and tents, with two solar panels to power the computers.

Hohmann arrived back at this place, on a June day in 2011, like a man very glad to be in the forest again after too many months deskbound in Leipzig. He was a robust 60-year-old, blue-eyed and bony, long conditioned to the steeplechase rigors of field primatology, and if I hadn't been pushing to stay at his heels, the six-hour hike would have taken me seven.

One morning I rose with the early crew, two lean young volunteers named Tim Lewis-Bale and Sonja Trautmann. We reached the bonobo nests at 5:20 a.m., before the drowsy animals began to stir. Their first act of the morning: a good piss. Lewis-Bale and Trautmann each stood beneath a nest tree, catching urine in a leaf. They pipetted this harvest into small vials, recorded the identity of each pisser, and then we were off on our morning chase.

That afternoon, Hohmann and I sat beneath one of the thatch roofs discussing bonobo behavior. Few other researchers have seen bonobos in the act of predation, and those few reports generally involved small prey such as anomalures (only at Wamba) or baby duikers. Animal protein, insofar as bonobos get any, seemed to come mainly from insects and millipedes. But Fruth and Hohmann reported nine cases of hunting by bonobos at Lomako, seven of which involved sizable duikers, usually grabbed by one bonobo, ripped apart at the belly while still alive, with the entrails eaten first, and the meat shared. More recently, here at Lui Kotale, they had seen another 21 successful predations, among which eight of the victims were mature duikers, one was a bush baby (a small primate that looks like a big-eyed squirrel), and three were monkeys. Bonobos preying on other primates: "This is a regular part of the bonobo diet," Hohmann said.

Sexiness, on the other hand, seemed to him less manifest than de Waal and other researchers had claimed. "I could show Frans some of the behaviors that he would not think are possible in bonobos," Hohmann said. Infrequent sex, for instance. Yes, there's a great diversity of sexual acts in the bonobo repertoire, but "a captive setting really amplifies all these behaviors. Bonobo behavior in the wild is different—must be different—because bonobos are very busy making their living, searching for food."

Hohmann mentioned other points of conventional wisdom against which he and Fruth dissented, including the notion that bonobo society is held together as a genial sisterhood by female bonding (they considered mother-son bonding at least as important) and the notion that bonobos aren't aggressive toward one another. Aggression may be rare and muted, he said, but that doesn't make it unimportant. Consider how subtle human aggression can be. Consider how a single violent act, or merely a mean one, can stick in a person's memory for years. "I think this is just what applies to bonobo behavior," he said. Life as a bonobo may be more stressful than it appears. Evidence of hidden anxieties had begun emerging from a hormone study by one of his postdocs, Martin Surbeck.

Analyzing fecal and urine samples, such as the ones Lewis-Bale and Trautmann gathered that morning, Surbeck found a surprising pattern: high levels of cortisol, a stress-related hormone, in some bonobo males. Cortisol levels were especially elevated among high-ranking males in the presence of estrous females. What did it imply? That a high-ranking bonobo male, walking a fine line between not enough machismo (which could cost him his status among males) and too much machismo (which could cost him his mating opportunities with imperious females), felt stressed by his complex situation. So it seemed that bonobos, though eschewing crude aggression and violence, are not carefree; they use sociosexual behaviors, diverse and relatively frequent, as a means of conflict management. "This is what makes them different," Hohmann said, "not that everything is peaceful."

THE BONOBO IS classified as endangered, and though protected by Congolese law, it continues to suffer from all-too-familiar problems, especially hunting for bush meat and habitat loss. Total bonobo numbers are even more difficult to estimate (let alone count) than for most large-bodied mammals because they are elusive and much of their forest habitat is so remote and unstudied. But current projections place the population at somewhere between 10,000 and 50,000, a broad range that reflects the uncertainty. Some of those are harbored within national parks and reserves, such as Salonga National Park and the Lomako-Yokokala Faunal Reserve. These "protected" areas may or may not provide effective security for bonobos and other wildlife, depending upon realities on the ground—for instance, whether guards have been hired and trained, paid their salaries, and supplied with adequate weapons to face poachers. The Democratic Republic of the Congo suffered severely from its seven decades of Belgian colonialism, followed by three decades of Mobutu's kleptocracy, followed by war; the context that frames all conservation efforts here is institutional dysfunction. Among the hostages to this situation is the bonobo, a species native to no country in the world except DRC. If it doesn't survive in the wild there, it will survive in the wild nowhere.

Two people who believe that it can survive are John and Terese Hart, conservationists who came originally to the Congo Basin in the early 1970s. Nowadays the Harts work with a young Congolese staff and a wide range of Congolese partners on a large project known as the TL2 Conservation Landscape, involving a region that straddles three of the big river's north-flowing tributaries in eastern Congo (the Tshuapa, Lomami, and Lualaba, hence TL2) and holds not just bonobos but also forest elephants, okapis, and a peculiar, newly discovered monkey called the lesula. Bonobos are still being poached in TL2, John told me, their carcasses often transported to market by bicycle to be sold as bush meat. With park status for part of TL2, antihunting regulations, support from local people, and enforcement at just a few checkpoints, he said, that trade could be choked off. TL2 has magnificent potential, but the constraints are formidable—even, I gathered, for such an irrepressible, experienced man as John Hart.

In Kinshasa I joined him and Terese, and we flew into Kindu, a provincial capital in eastern Congo (and a jumping-off point to TL2) on the west bank of the Lualaba River, which defines the eastern limit of bonobo distribution. In Kindu, after days of bureaucratic delay exacerbated by a local grudge, we finally got approval for a little five-day expedition through TL2. Around 4:00 p.m.—late for a departure, but we were concerned not to lose another day—we climbed into a large dugout canoe, before the officials could change their minds. We were joined by two of the Harts' trusted Congolese colleagues, plus a visiting biologist, plus a colonel and a soldier (both with Kalashnikovs) as our military escort. There was also a man from the Immigration directorate, assigned at the last minute to shadow us. The man from Immigration, who must have been tossed into this assignment by his boss on short notice, wore street shoes and carried his change of shirt in a briefcase. We'll be out about 30 days, and you'll need to help us kill crocodiles for food, John teased him, as the outboard pushed us weakly away from Kindu, and we set our course midstream down the Lualaba.

The river was brown, flat, and a thousand yards wide. The sun, sinking low behind the dry-season haze, looked like a great bloody yolk. I watched a pair of palm-nut vultures pass overhead and then, to the east, a flock of fruit bats circling their roost. Dusk faded quickly to dark, and the river glowed sepia with reflection from a waxing crescent moon. The air cooled;

we pulled on jackets. Hours later we grounded at a village on the left bank that marked our trailhead for this hike into bonobo country.

It had to be the left bank, I knew. Geography is fate. There were no bonobos, anywhere, on the right.

· ·

The factors that menaced bonobos most severely in 2013, when this piece first appeared, continue to menace bonobos today: illegal hunting for bush meat and the destruction and degradation of their habitat. What drives those factors are population growth, poverty on rural landscapes, and hunger; what exacerbates them are corruption within institutions and profiteering by opportunists in the middle. Poor people continue to suffer, in the areas of interface between village and forest, and wild animals do too. The DRC now contains about 95 million humans and—a narrowed range, as estimate methods have improved—probably between 10,000 and 20,000 bonobos. Whatever the actual number, somewhere within that range or just outside, the safe assumption is that they are fewer than they were a decade ago. And this little-known, little-protected creature, confined to an area beneath the great northern bend of the Congo River, stretching southward to the Sankuru and eastward to the Lualaba, is one of humanity's two closest living relatives.

There is some good news. On July 7, 2016, the government of DRC designated a sizable portion (almost 900,000 hectares, or 3,500 square miles) of the TL2 conservation area, at that easternmost edge of bonobo distribution, to be Lomami National Park. It became the country's first new national park since 1970, and if management, antipoaching enforcement, and socioeconomic relations with the people (Lengola, Mbole, and five other ethnic groups) of the surrounding buffer zone are good (see "People's Park," pages 275–294, for a paradigm of how that might be done), then Lomami National Park will help protect not just its bonobos and other endangered and vulnerable species such as okapi, forest elephant, and Congo peafowl, but also populations of 12 other primate species. How many places on Earth—how many contiguous areas of similar size—harbor 13 species of nonhuman primate? The answer has got to be "few, if any," and you would need a biogeographic map of Brazil even to begin addressing the question. The 13 in Lomami National Park include the lesula (Cercopithecus lomamiensis), a species of guenon unknown to science until 2007, when a team led by John and Terese Hart (in conjunction with the Lukuru Foundation) recognized its uniqueness during a reconnaissance trip by dugout canoe up the Lomami River. As for the bonobos, their range extends beyond the Lomami, almost to the left bank of the Lualaba. Down there near the Congo headwaters, as elsewhere, they are the left bank ape.

Cross the Lualaba, headed east toward Africa's Great Rift Valley, and you reenter a relatively impoverished world: one without bonobos.

THE SHORT HAPPY LIFE
OF A SERENGETI LION

·⊹·⊹·⊹·

C-Boy Against the Odds

·······································

*Nick is going to do the Serengeti lion, the editors said to me.
Are you in?*

I'm definitely in, I said.

Stories for National Geographic *often begin with the photography side. That was usually fine with me, because it gave me the opportunity to work with some of the world's best and most committed photographers, and it saved me the trouble of pitching ideas and pushing them through editorial committees. I can't recall whether it was Ollie Payne, my text editor for so many stories, who first told me that Nick Nichols was arranging a major effort in Serengeti, or Kathy Moran, the dedicated photo editor whose stable of galloping horses included Nick. But that didn't matter, not after I had spent a dozen years working for the magazine and living through so many field days and nights of effort, misadventure, exhilaration, and well-attuned partnership with Nick. There are two people in the world to whom, if they say, "We're going to parachute into Borneo, live on crackers for a month, and look for the world's largest spider; we have your chute packed and an extra crate of crackers; we leave Tuesday. Are you in?" my answer is "I'm in." (Although I might ask: "How large a spider?") Nick Nichols is one of those two people.*

Besides, who doesn't want to spend weeks living out of a comfortable safari tent in the backcountry of Serengeti National Park, following lions with one of the world's leading lion biologists and his stalwart field assistants, not to mention a trusted photographer friend?

It was an irresistible prospect, and it played out as well as expected. But the most affecting part of this experience was encountering C-Boy, a very special lion, a creature of beauty and force and gravitas. And the most amazing part was that someone paid me to do it.

A note on the title: Obviously it's an allusion to Hemingway's story, "The Short Happy Life of Francis Macomber." The pointed irony of Hemingway's title, which depends on the sly omission of a comma between "Short" and "Happy," is that the protagonist enjoys a very short period of happiness (spoiler alert in case you haven't read the story) before his otherwise unhappy life ends abruptly. Stated otherwise: Insert a comma, as conventionally expected, and the adjective "Short" modifies "Life." Omit the comma, as Hemingway did, and suddenly "Short" modifies "Happy Life."

Lions are involved, too, in Hemingway's version. I wanted readers to see, or at least to sense subliminally, if they knew and remembered the story, that I had echoed Hemingway but turned the tables as to who was the protagonist. For that connotation to work, it was necessary to preserve the original, eloquent omission of the comma.

Copy editors at National Geographic *believe they know a thing or two about grammar, and I honor that. I venerate editors who are good, including good copy editors. But the guiding wisdom of copy editor grammar says that, when you have two sequential adjectives before a noun, you separate them with a comma. "I spent a short, happy vacation at the beach."*

Copyediting by the rules is sometimes oblivious to nuance.

It would have embarrassed me to throw an authorial tantrum over this point, but I did say: If you insist on adding the comma, let's trashcan that title and I'll give you a different one. To run it with the comma would ruin the effect and be nonsensical.

Fussy writers, we live in our solitary rooms, trying to send messages to you readers in corked bottles tossed upon the waves. But we have friends. The heroes of this little anecdote were two friends, and working partners, of mine. Nick Nichols stood up for me. He was

there in the editorial offices while I was not. You should give David
his title, he argued to Editor in Chief Chris Johns. Chris did.

· ·

They say that cats have nine lives, but they don't say that about the Serengeti lion. Life is hard and precarious on this unforgiving landscape, and dead is dead. For the greatest of African predators as well as for their prey, life spans tend to be short, more often terminating abruptly than in graceful decline. An adult male lion, if he's lucky and durable, might attain the advanced age of 12 in the wild. Adult females can live longer, even to 19. Life expectancy at birth is much lower, for any lion, if you consider the high mortality among cubs, half of which die before age two. But surviving to adulthood is no guarantee of a peaceful demise. For a certain young male, black-maned and robust, known to researchers as C-Boy, the end seemed to have arrived on the morning of August 17, 2009.

A Swedish woman named Ingela Jansson, working as a field assistant on a long-term lion study, was there to see it. She knew C-Boy from previous encounters; in fact, she had named him. (By her recollection, she had "boringly" labeled a trio of new lions alphabetically, as A-Boy, B-Boy, and C-Boy.) Now he was four or five years old, just entering his prime. Jansson sat in a Land Rover, 30 feet away, while three other males ganged up on C-Boy and tried to kill him. His struggle to survive against those daunting odds, dramatic and immediate, reflected a larger truth about the Serengeti lion: Continual risk of death, even more than the ability to cause it, is what shapes the social behavior of this ferocious but ever jeopardized animal.

On the day in question, near the dry bed of the Seronera River, in the heart of Serengeti National Park, Jansson had come to check on a pride known as Jua Kali. A pride consists of females with young. Jansson was also alert for adult males, including those "resident" with the pride. Male lions, not strictly belonging to any pride, instead form coalitions with other males and exert controlling interest over one or more prides, fathering the cubs and becoming resident, loosely associated with the pride. They also play an important role in helping kill prey—especially with larger and more dangerous animals, such as Cape buffalo or hippos—thereby contributing something besides sperm and protection to the life of the pride. The resident males of Jua Kali, Jansson knew, were C-Boy and his sole coalition partner, a golden-maned lothario named Hildur.

Approaching the river, she saw in the distance one male chasing another. The fleeing lion was Hildur. Fleeing from what, and why, she didn't at first understand. Then she found a group of four males in the grass. They had settled themselves in a squarish pattern, each about five strides away from the others. She recognized them—some of them, anyway—as members of another coalition, a group of four ambitious young adult males, notorious in her record cards as the Killers.

One lion had a bloody tooth, the lower right canine, suggesting a very recent fight. Another was hunkered flat, as though wishing he could disappear into the ground. From the flattened male came a steady, nervous growl. Driving closer, Jansson saw the dark tinge of his mane and realized this was C-Boy, wounded, isolated, and surrounded by three of the Killers.

She had also noticed a lactating female, the radio-collared lioness of the Jua Kali pride. Lactation meant young cubs, hidden somewhere in a den, the presumptive father of which was C-Boy or Hildur. So this standoff between C-Boy and the Killers was more than a pointless rumble. It was a challenge for controlling rights to a pride. If the new males took over, they would kill the young of their rivals to bring the females quickly back into heat.

Seconds later, the fight erupted again. The three Killers circled C-Boy and took turns lunging at him from behind, lashing into his haunches, biting at his spine, as he spun and snarled and rolled desperately to escape. Close enough almost to feel the spray of spit, to smell the malice, Jansson gaped from her car window, taking photos. Dust flew, C-Boy whirled and roared, and the Killers played their advantage, avoiding his jaws, backing off, coming at him again from the rear, sinking their teeth, scoring hurts, until the hide of his hindquarters looked like a perforated old pelt. Jansson thought she was witnessing the terminal event of a lion's life. If the immediate injuries didn't kill him, she reckoned, the later bacterial infections would.

Then it was over, as abruptly as it had begun. Maybe a minute of fighting. They separated. The Killers strolled off and positioned themselves atop a termite mound, with a commanding view of the river, while C-Boy slunk away. He was alive, for the moment, but defeated.

Jansson didn't see him for two months. He might have been dead, she guessed, or at least debilitated. In the meantime, the Killers began having their way with the Jua Kali females. The small cubs of C-Boy's or Hildur's paternity disappeared—killed by the conquering males, or maybe abandoned to starvation, or neglected just enough to get eaten by hyenas. The females would go back into estrus now, and the Killers would father new litters. C-Boy was yesterday's favorite, yesterday's stud. The Jua Kalis would forget him. This is the cold arithmetic of lion society.

TIGERS ARE SOLITARY. Cougars are solitary. No leopard wants to associate with a bunch of other leopards. The lion is the only felid that's truly social, living in prides and coalitions, the size and dynamics of which are determined by an intricate balance of evolutionary costs and benefits.

Why has social behavior, lacking in other cats, become so important in this one? Is it a necessary adaptation for hunting large prey such as wildebeests? Does it facilitate the defense of young cubs? Has it arisen from the imperatives of competing for territory? As details of leonine sociality have emerged, mostly over the past 40 years, many of the key revelations have come from a continuous study of lions within a single ecosystem: the Serengeti.

Serengeti National Park encompasses 5,700 square miles of grassy plains and woodlands near the northern border of Tanzania. The park had its origin as a smaller game reserve under the British colonial government in the 1920s and was established formally in 1951. The greater ecosystem, within which vast herds of wildebeest, zebra, and gazelle migrate seasonally, following the rains to fresh grass, includes several game reserves (designated for hunting) along the park's western edge, other lands under mixed management regimes (including the Ngorongoro Conservation Area) along the east, and a transboundary extension (the Masai Mara National Reserve) in Kenya. In addition to the migratory herds, there are populations of hartebeests, topi, reedbucks, waterbuck, eland, impalas, buffalo, warthogs, and other herbivores living less peripatetic lives. Nowhere else in Africa supports quite such a concentrated abundance of hoofed meat, amid such open landscape, and therefore the Serengeti is a glorious place for lions. It's also an ideal site for lion researchers.

George Schaller arrived in 1966, by invitation of the director of Tanzania National Parks, to study the effects of lion predation on prey populations—and to learn as much as he could in the process about the dynamics of the entire ecosystem. Schaller, a legendarily tough and astute field biologist, had earlier done pioneering research on mountain gorillas. If you're making the first detailed study of any species, he told me about that experience, "you grab what you can." He grabbed a cornucopia of data during just over three years of intensive fieldwork, and his subsequent book, *The Serengeti Lion*, became the foundational text.

Other researchers followed. A young Englishman named Brian Bertram succeeded Schaller and stayed four years, long enough to begin teasing out the social factors that affect reproductive success and to explain an important phenomenon: male infanticide. Bertram documented four cases (with many others suspected) in which a new coalition of males killed cubs of a pride it had just taken over. Jeannette Hanby and David Bygott came next and assembled evidence that forming coalitions—especially coalitions of three or more—helps male lions gain and hold control of prides and thereby produce more surviving offspring. Hanby and Bygott studied some of the same prides in the same areas as Bertram and Schaller had.

Then, in 1978, Craig Packer and Anne Pusey took over the study, after having done fieldwork at the Gombe Stream Research Center (also in Tanzania) with Jane Goodall. Pusey stayed with the lion project a dozen years, co-authoring some important papers, and Packer remained for decades more, leading the Serengeti Lion Project, of which Ingela Jansson's work was part. Packer became the world's leading authority on the behavior and ecology of the African lion. With Packer's 35 years of work added to what Schaller and the others did, the Serengeti Lion Project represents one of the world's longest continuous field studies of a species. Such continuity is especially valuable, allowing scientists to view events in broad context and distinguish the transitory from the essential. "If you have a long-term data set," Schaller told me, "you find out what actually happens."

One thing that happens is death. Although it's ineluctable for every creature, the particulars of its timing and its cause add up to patterns that matter.

AFTER HIS HARROWING experience with the Killers, C-Boy surrendered his claim on the Jua Kali pride and shifted his attentions east. Hildur, his coalition partner, who had been so little help in the pinch, went with him. By the time I got a glimpse of C-Boy three years later, he and Hildur had established control over two other prides, Simba East and Vumbi, whose territories lay amid the open plains and kopjes (rocky outcrops) south of the Ngare Nanyuki River. This was not the most hospitable part of the Serengeti for lions and their prey—during the dry season it could be lean and difficult—but it offered C-Boy and Hildur an opportunity to start fresh.

I was traveling through that area with Daniel Rosengren, another adventuresome Swede, who had taken over the lion-monitoring role from Jansson. Way out here, east of the main tourism area and south of the river, the great vistas of grassland rise and fall smoothly, like oceanic swells, punctuated every few miles by a cluster of kopjes. The kopjes, granitic lumps festooned with trees and shrubs, standing above the plains like garnished gumdrops, offer shade and security and lookout points for resting lions. You could drive for days in this corner of the park and not see a tourist vehicle. Along with Nick Nichols and his photo team, who were spending months at a field camp near the riverbed, we had the area to ourselves.

That afternoon the radio signal in Rosengren's headphones led us to Zebra Kopjes, where, amid the cover, we found the collared female of the Vumbis. Beside her was a magnificent male with a thick mane that cascaded off his neck and shoulders like a velvet cape, shading from umber to black. It was C-Boy.

From just 40 feet away, even through binoculars, I could detect no sign of injuries to his flanks or rear. The punctures had healed. "On lions," Rosengren told me, "most scars disappear after a while, unless they're around the nose or mouth." C-Boy had made a new life for himself in a new place, with new lionesses, and seemed to be thriving. He and Hildur had fathered several more litters of cubs. And just the night before—so we heard from Nichols, who had seen it—the Vumbi females brought down an eland, a very large hunk of prey, after which C-Boy had laid his imperious male forepaw on the carcass, claiming first bites. C-Boy had fed on the eland alone, taking choice morsels but not too much, before allowing the lionesses and their cubs to get at it. Hildur had been elsewhere, presumably consorting with another estrous female. They were living the good life, those two, with all the prerogatives of resident male lions. This was just 12 hours before we saw evidence suggesting that trouble had followed them east.

The trouble was male competition. Early next morning Rosengren drove us north from Nichols's camp to the river, seeking a pride known as Kibumbu, whose small cubs had been fathered by still another coalition. Those males had gone absent in recent months—departed to places unknown, for reasons unknown—and Rosengren wondered who might have supplanted them. That was his assignment, within the broader context of Packer's lion studies: to chronicle the comings and goings, the births and deaths, the affiliations and retreats that affect pride size and territorial tenure. If the Kibumbus had new daddies, who might they be? Rosengren had a suspicion—and it was confirmed when, amid the high grass of the riverbank, we came upon the Killers.

They were handsome devils, a quartet of eight-year-old males, resting in a companionable cluster. They looked forbidding and smug. They're probably two sets of brothers, Rosengren told me, born within months of each other in 2004. They had been dubbed "the Killers" by another field assistant back in 2008, based on his inference that they had killed three collared females, one by one, rather systematically, in a drainage just west of the Seronera River. Such male-on-female violence wasn't utterly aberrant—it might even be adaptive for males in some cases, opening space for prides that they control by removing competition in the form of neighboring females—but in this case it won the males a malign reputation.

Although Rosengren told me their individual names as recorded on the cards (Malin, Viking, et cetera), his preference was to call them by their numbers: 99, 98, 94, 93. Those numerals did seem somehow more concordant with their air of opaque, stolid menace. Male 99, seen in profile, had the convex nose line of a Roman senator, as well as a darkish mane, though not so dark as C-Boy's. Inspecting him through binoculars, I noticed a couple of small wounds on the left side of 99's face.

Rosengren eased the Land Rover closer, and two of the others, 93 and 94, stirred, turning toward us. In the golden light of sunrise, we saw facial injuries on them too: a slice to the nose, a bit of swelling, a gash below the right ear still glistening with pus. Those are fresh, Rosengren said. Something happened last night. And not just a spat over shared food; coalition partners don't do such damage to one another. It must have been a brawl with other lions. That raised two questions: Whom had the Killers fought? And what did the other guy look like this morning?

Then, as the day progressed and we made other rounds, it seemed that C-Boy was missing.

"MOSTLY LIONS DIE because they kill each other," Craig Packer told me, in response to a question about fatalities. "The number one cause of death for lions, in an undisturbed environment, is other lions."

He broke that into categories: At least 25 percent of cub loss results from infanticide by incoming males. Females too, given the chance, will sometimes kill cubs from neighboring prides. They will even kill another adult female, he said, if she unwisely wanders into their ambit. Resources are limited, prides are territorial, and "it's a tough 'hood out there."

Males operate just as jealously. "Male coalitions are gangs, and if they find a strange male that's hitting on their ladies, they'll kill him." And males will kill adult females if it suits their purposes, as the Killers had shown. You see a lot of bite wounds on lions, reflecting the competitive struggle for food, territory, reproductive success, sheer survival. With luck, the wounds heal. Less luck, and the loser is killed in a fierce leonine battle, or he limps away, losing blood, maybe crippled, maybe destined to die slowly of infection or starvation. "So the lion is the number one enemy of lions," Packer said. "It's why, ultimately, lions live in groups." Holding territory is crucial, and the best territorial locations—places he calls hot spots, such as stream confluences, where prey tend to become concentrated—serve as incentive for social cooperation. "The only way you can monopolize one of those very valuable and very scarce hot spots," he said, thinking like a lion, is as "a gang of like-sexed companions who work as a unit."

That theme emerged strongly from Packer's research, done with various collaborators and students over the decades. It's not just the need for joint effort in making and defending kills, he found, that drives lionesses to live in prides. It's also the need to protect offspring and retain those premium territories. His data showed that, although pride size varies widely, from just one adult female to as many as 18, prides in the middle range succeed best at protecting their cubs and maintaining their territorial tenure. Prides that are too small tend to lose cubs. Periods of estrus for the adult females often are synchronized—especially if an episode of male infanticide has killed off all their young and reset their clocks—so that cubs of different mothers are born at about the same time. This allows the formation of crèches, lion nursing groups in which females suckle and protect not just their own cubs but others too. Such cooperative mothering is further encouraged by the fact that the females of a pride are related—as mothers and daughters and sisters and aunts, sharing a genetic interest in one another's reproductive success. But prides that are too large do poorly also, because of excessive within-pride competition. A pride of two to six adult females seems to be optimal on the plains.

Male coalition size is governed by similar logic. Coalitions are formed, typically, among young males who have outgrown the natal pride and gone off together to cope with adulthood. One pair of brothers may team with another pair, their half siblings or cousins, or even with unrelated individuals that turn up, solitary, nomadic, and needing partnership. Put too many such males together as a roving posse, each hungry for food and chances to mate, and you have craziness. But a lone male, or a coalition that's too small—just a pair, say—will suffer disadvantages also.

That was C-Boy's dilemma. With no partner other than Hildur, a handsome enough male who showed great eagerness to mate but little to fight, C-Boy confronted the Killers, in their aggressive ascendancy, virtually alone. Not even his resplendent black mane could neutralize four-against-one odds. Maybe by now he was already dead. If so, Rosengren and I realized, those minor battle injuries on the faces of the Killers might be the last evidence of C-Boy that anyone would ever see.

THAT NIGHT THE Killers made another move into new territory. They had rested all day by the riverbank, letting the sun cook their faces and dry their sores. About two hours after sunset, they started roaring. Their joined voices broadcast a message of some sort—maybe, *Here we come!*—into the distance. Then they set out, all four together, on what looked like a purposeful march. Rosengren and I got the word by walkie-talkie from Nichols, who had been keeping vigil. We jumped into Rosengren's Land Rover and headed out through the blackness, beginning what I recall as the Night of the Long Follow.

Converging with Nick's vehicle, we climbed in and stayed with the lions—five of us now, Nick's wife, Reba Peck, at the wheel, easing along, headlights dimmed. There was no moon. Nick had night vision goggles and an infrared camera. His assistant and videographer, Nathan Williamson, sat ready to capture sound or deploy the infrared floodlights. We were a journalistic gunship, bristling with armaments, rolling slowly along behind the lions. They showed no concern whatsoever about our presence. They had other things in mind.

We followed them up an old buffalo track, then through a tight grove of fever trees, Reba coaxing the vehicle patiently around aardvark holes, over crunching thorn branches, across a sumpy stream bottom. Please don't get stuck, we all thought. With four Killers nearby, nobody wanted to climb out and push. We didn't get stuck. The lions walked in single file, steady, unhurried, neither waiting for us nor trying to lose us. We kept them in view with the low headlights and, where those didn't reach, a monocular thermal scope. Through the scope, as I sat atop the Rover's jouncing roof, I saw four lion bodies glowing like candles in a cave.

Suddenly another large figure swung up alongside us, its eyes shining orange when I swept it with my headlamp. It was a lioness, making herself known to the Killers. Rosengren couldn't recognize her, in this fleeting glimpse, but presumably she was in heat. So she was taking a sex-mad risk, probably larger than she could guess, given the record of these males. When they noticed her, and wheeled toward her, she ran off coyly, pursued by all four, and for a moment we thought we had lost them. But only one male stayed on her tail; we wouldn't see him again all night. The other three reassembled themselves, after this flirty distraction, and continued their march.

They crossed a dirt two-track (the main east-west "road," which we used coming and going to camp) and angled south, now brazenly entering the territory of the Vumbi pride and its resident defenders, C-Boy and Hildur. They paused here and there to scent-mark, rubbing their foreheads against bushes, scratching and spraying the ground. This wasn't a sneak attack; they were advertising themselves, making a statement. Too bad, Rosengren noted, that we don't have some sort of fancy scope to illuminate those smells.

By now they had turned and were headed toward Nick's camp, so Williamson radioed ahead and warned the kitchen crew to stay in their tents. But the three lions didn't care about our little canvas compound, with its odors of popcorn and chicken and coffee, any more than they cared about us; about a quarter mile short, they bedded down to rest. During this hiatus, just before midnight, Nick and his team went back to camp. Before going, they retraced our path back to the other vehicle, dropping Rosengren and me there so he and I could retrieve it and rejoin the Killers. He took the first sleeping shift, snoring gently in the back of the Land Rover, while I sat up, keeping watch. Half an hour later the lions stood and began moving again. I woke Rosengren, and we followed.

And that's how it went—a stretch of them walking and us driving, then a stretch of sleeping, Rosengren and I trading duties—for the rest of the night. Occasionally, during a stop, they let their voices rise in another chorus of roars. The roar of three lions heard at close range, let me tell you, is an imposing sound: high in decibels but throaty and rough, as though scraped up from a deep iron bin of primordial power and confidence and threat. No one answered these calls. In the wee hours, the trio met a lone Thomson's gazelle; that poor gazelle must have been terrified, but as the lions made a perfunctory try, it bounded safely away. One Tommy, divided three ways, was scarcely worth the trouble. As dawn came, they were back on the road after their big loop through Vumbi territory, strolling casually west toward a familiar kopje where they would find shade for the day. It was Saturday morning. Rosengren and I left them there.

The wounds on their faces, and the absence of C-Boy, were still unexplained. Lion politics along the Ngare Nanyuki River seemed to be in flux.

Late Saturday afternoon, we found the Vumbi pride at Zebra Kopjes, a couple miles south of where the Killers had made that intrusive circuit. Maybe the pride had been driven down there by the minatory roaring, or maybe they had just wandered. We counted three females, resting contentedly amid the shaded lobes of granite, and all eight cubs. Another female, we knew, was off on a mating foray with lover boy Hildur. No sign of C-Boy. His absence seemed slightly ominous.

Sunday afternoon, back to Zebra Kopjes. Hildur and his female had rejoined the group, but not C-Boy. Let's try Gol Kopjes, Rosengren suggested. With luck we'll see the Simba East pride, and he might be with them. Yes, I said; that's my priority, I want to find him, dead or alive. So we drove southwest, rising and descending gently across the swales of grassland, while Rosengren listened in his headphones for the bleeping radio collars of Simba East. At a small kopje near the main Gols we located them: three females and three large cubs, lounging amid the radiant rocks. But again, no sign of C-Boy.

Rosengren, at this point, admitted to some worry. His job was not to root for favorites, of course, but to monitor events, including the natural phenomena of lion-on-lion violence and pride takeover, but he had his sympathies. It's beginning to seem, he said sadly, that C-Boy must have fallen victim to the Killers.

With a lavender Serengeti sunset painting the horizon behind us, we drove back to Zebra Kopjes. Nick and Reba were still there, with the Vumbis, who had hunkered together in the grass and begun roaring—one voice, then another, then three together, rumbling out across the plains beneath a now darkening sky and a small waxing crescent of moon. Lion roars can carry a range of meanings, and this chorus bore a mysterious, lonely tone. When they fell silent, we listened with them. No response.

Nick and Reba departed for camp. Rosengren circled our vehicle into a spot just beside the reclining Vumbis. He wanted me to experience the fearsome thrill of taking lion roars point-blank in the face. This time Hildur joined in, his deep male basso rasping and thundering, almost shaking the car. Once they finished, we again listened intently. And again nothing. Now I was ready to leave. For journalistic purposes, I was prepared to list C-Boy as "missing, suspected dead."

Wait, Rosengren said. There was a scuffle in the darkness around us. Give me your headlamp, he said. Swinging the beam from left to right, across Hildur and the others, Rosengren brought it to rest on a new figure, a large one, with a very dark mane: C-Boy. He was back. He had come running to the sound of their roars.

His face was smooth. His flanks and buttocks were intact. Whomever the Killers had mugged two nights ago, it wasn't him. Or else he had bested them without taking a scratch. He settled comfortably beside the collared female. Soon he would be mating again. He was an eight-year-old lion, healthy and formidable, commanding respect within a pride.

It was all very temporary. C-Boy's life might stretch forward a few years, beyond this moment, into infirmity, injury, mayhem, displacement, starvation, and death. The Serengeti offers no mercy to the elderly, unlucky, or impaired. He wouldn't always be happy. But he looked happy now.

· ·

Five years passed before I heard further news of C-Boy, and when I did hear, the report was sad but inevitable, and it left his dignity unmarred. For this, see "Elegy for a Lion," pages 211–214.

I have retained the chronological order in which these three lion-related stories first appeared because I hope that the next chapter, "Tooth and Claw," though it doesn't mention him individually, will help you further understand the world of perils within which C-Boy lived. It may also help you appreciate his extraordinary good fortune to have prowled out his days within the protected landscape of Serengeti National Park.

Lions, in Tanzania and Kenya and other countries of sub-Saharan Africa, occupy habitat outside of park boundaries as well as within them. They share borderlands between savannas and crop fields. Sometimes they follow prey, such as bush pigs, into those crops—amid which a farmer might be standing vigil at night, with no more weapon than a big spoon to bang on a metal pot, in defense of his or her ripening maize. And lions in human habitat may prey upon livestock as well as on native ungulates. The following story tries to make these realities concrete.

It's one thing to observe a magnificent lion from the safety of a Land Rover, as researchers and journalists and tourists do. It's quite another thing to walk the thorn scrub or the savanna with just a spear or a hoe in your hand, idling along with your cows or goats, in acute awareness that lion eyes may be watching as you invade what those felines are entitled to consider their *space.*

· ·

TOOTH
AND CLAW

·❦·❦·❦·

For People and Lions,
Living Together Is the Hard Part

Lions of the Serengeti are regal. Lions of the borderlands outside of national parks, on which people as well as wildlife live and seek their sustenance, are highly armed, highly dangerous, and highly vulnerable participants in an intricate scrum.

Chris Johns, at the time editor in chief of National Geographic magazine, with long experience himself as a photographer of African wildlife, wisely saw that it would be right to include, in the same issue as our Serengeti story, a separate treatment of the very raw human realities of being a local person, a farmer or herder of livestock, living in lion habitat.

The assigned photographer for that story was Brent Stirton, a South African, doughty in tough situations, humane in delicate ones, with whom I would have the privilege of working again. For this brief story, which was originally conceived as something broader, I traveled from the Serengeti to the Selous Game Reserve in southern Tanzania, up to the Kenyan borderland plains outside Amboseli National Park, and to Benin in West Africa, along the northern boundary of which live some of the last lions in West Africa. Most of that didn't make it into the piece, but I learned a lot about how hard it is to be a human, or a lion, on some of Earth's more austere landscapes.

L ions are complicated creatures, magnificent at a distance yet fearsomely inconvenient to the rural peoples whose fate is to live among
them. They are lords of the wild savanna but inimical to pastoralism
and incompatible with farming. So it's no wonder their fortunes have
trended downward for as long as human civilization has been trending up.

There is evidence across at least three continents of lions' glory days
and their decline. Chauvet Cave, in southern France, filled with vivid
Paleolithic paintings of wildlife, shows us that lions inhabited Europe along
with humans 30 millennia ago; the Book of Daniel suggests that lions
lurked at the outskirts of Babylon in the sixth century B.C.; and reports of
lions surviving in Syria, Turkey, Iraq, and Iran continued until well into
the 19th and 20th centuries. Africa alone, during this long ebb, remained
the reliable heartland.

But that has changed too. Surveys and estimates suggest that lions have
disappeared from about 80 percent of their African range. No one knows
how many lions survive today in Africa—as many as 35,000?—because
wild lions are difficult to count. Experts agree, though, that just within
recent decades, the overall total has declined significantly. The causes are
multiple—including habitat loss and fragmentation, poaching of lion prey
for bush meat, poachers' snares that catch lions instead, displacement of
lion prey by livestock, disease, spearing or poisoning of lions in retaliation
for livestock losses and attacks upon humans, ritual killing of lions (notably
within the Maasai tradition), and unsustainable trophy hunting for lions,
chiefly by affluent Americans.

Assessments compiled by scientists from Panthera (an international
felid conservation group), Duke University, the National Geographic
Society's Big Cats Initiative, and others have indicated that African lions
live in scores of distinct areas (as many as 67), the largest and most secure
of which can be considered strongholds. But the smallest contain only tiny
populations, isolated, genetically limited, and lacking viability for the long
term. In other words, African lions inhabit an archipelago of insular refuges,
and the smallest of those marooned populations face continuing peril of
extinction.

WHAT CAN BE done to stanch the losses and reverse the trend? Some experts say we should focus efforts on the strongholds, such as the Serengeti eco-system (spanning from northern Tanzania into southwestern Kenya), the Selous ecosystem (southeastern Tanzania), Ruaha-Rungwa (western Tanzania), Okavango-Hwange (Botswana into Zimbabwe), and Greater Limpopo (at the shared corners of Mozambique, Zimbabwe, and South Africa, including Kruger National Park). Those five ecosystems alone account for roughly half of Africa's lions, and each contains a genetically viable population. The veteran lion biologist Craig Packer (see "The Short Happy Life of a Serengeti Lion," pages 187–201) once offered a drastic suggestion for further protecting some strongholds: Fence them, or at least some of their margins. Investing conservation dollars in chain-link and posts, combined with adequate levels of patrolling and repair, Packer argued, is the best way to limit illegal entry into protected areas by herders, their livestock, and poachers, as well as reckless exit from those areas by lions.

Other experts strongly disagree. In fact, this fencing idea went against three decades of conservation theory, which stresses the importance of connectedness among habitat patches. Packer knew that, and even he wouldn't put a fence across any valuable route of wildlife dispersal or migration. But consider, for instance, the western boundary of the Serengeti ecosystem, where the Maswa Game Reserve meets the agricultural lands of the Sukuma people beyond. If you flew over that area at low elevation, around the time Packer made his suggestion, you saw the boundary as a stark edge, delineated by the slash of a red clay road. East of it lay the rolling green terrain of Maswa, covered with acacia woodlands and lush savanna, a virtual extension of Serengeti National Park. West of the road, in the Sukuma zone, with its large human population, you looked down on mile after mile of cotton fields, cornfields, teams of oxen plowing bare dirt, paddies, and brown-and-white cows standing in pens. A fence along that boundary, as Packer asserted, could do no harm and possibly some good. It was a special case, but it was enough to open a heated discussion.

Trophy hunting is also controversial. Does it contribute to population declines because of irresponsible overharvesting? Or does it effectively monetize the lion, bringing cash into local and national economies and providing an incentive for habitat protection and sustainable long-term management? The answer depends—on particulars of place, on which lions are targeted (old males or young ones), and on the integrity of management, both by the hunting operator and the national wildlife agency. Certainly, there are abuses—countries in which hunting concessions are granted corruptly, situations in which little or no hunting income reaches the local people who pay the real costs of living amid lions, concession lands on which too many lions are killed. But in places such as the Maswa Game Reserve—where hunts are scrupulously managed in cooperation with the Friedkin Conservation Fund, an organization that cares more about habitat protection than about revenue—the effect of a ban on hunting could be perverse.

Hunting of captive-bred lions released into fenced areas on private ranches, as widely practiced in South Africa, raises a whole different set of questions. In a recent year, 174 such lion-breeding ranches operated in the country, with a combined stock of more than 3,500 lions.

Proponents argue that this industry may contribute to lion conservation by diverting trophy-hunt pressure from wild populations and by maintaining genetic diversity that could be needed later. Others fear it may undercut the economics of lion management in, say, Tanzania, by offering cheaper and easier ways to put a lion head on your rec-room wall.

And then there's the matter of what happens to the rest of the lion. The export of lion bones from South Africa to Asia, where they are sold as an alternative to tiger bones for supposed medicinal and tonic purposes, constitutes a dangerous trend that surely increases demand.

Bottom line: Lion conservation is an intricate enterprise that must now reach across borders, oceans, and disciplines to confront a global market in an abstract but saleable commodity—dreams of the wild.

BUT CONSERVATION BEGINS at home, among people for whom the sublime and terrifying wildness of a lion is no dream. One set of such people are the Maasai who inhabit group ranches bordering Amboseli National Park, on the thornbush plains of southern Kenya. Since 2007, a program there called Lion Guardians has recruited Maasai warriors—young men for whom lion killing has traditionally been part of a rite of passage, known as *olamayio*—to serve instead as lion protectors. These men, paid salaries, trained in radiotelemetry and GPS use, track lions daily and prevent lion attacks on livestock. The program, small but astute, seems to be succeeding. Lion killings have decreased, and being a Lion Guardian is now a prestigious role within those communities.

I spent a day with a Lion Guardian named Kamunu, roughly 30 years old, serious and steady, whose dark face tapered to a narrow chin and whose eyes seemed permanently squinted against sentiment and delusion. He wore a beaded necklace, beaded earrings, and a red shuka wrapped around him; a Maasai dagger was sheathed to his belt on one side, a cell phone on the other. Kamunu had personally killed five lions, he told me, all for olamayio, but he didn't intend to kill any more. He had learned that lions could be more valuable alive—in money from tourism, wages from Lion Guardians, and the food and education such cash could buy for a man's family.

We walked a long circuit that very hot day, winding through acacia bush, crossing a dry riverbed, Kamunu following lion spoor in the dust and me following him. Probably we traipsed about 16 miles. In the morning we tracked a lone adult, recognizable to Kamunu from its big pugmarks as a certain problematic male. When we met a sizable file of cows headed for water, their bells clanking, Kamunu warned the several Maasai boys attending the cattle to stay clear of that lion.

Around midday he picked up a different trail, very fresh, left by a female with two cubs. We saw her flattened day bed in the herbage beneath a bush. We traced her sinuous route into a grove of scrubby myrrh trees that grew thicker as we went. Kamunu moved quietly. Finally, we stopped. I saw nothing but vegetation and dirt.

They're very close, he explained. This is a good spot. No livestock nearby. We don't want to push any closer. We don't want to disturb them. No, we don't, I agreed.

"We think they are safe here," he told me. It's more than can be said for many African lions, but at that moment, in that place, it was enough.

The organization Lion Guardians continues its work, in the Amboseli ecosystem and elsewhere, and the tensions between lions and people continue too—mitigated in some places by the sort of community-level efforts that Lion Guardians supports, exacerbated in others by the classic old drivers of conflict: human population growth, landscape conversion, illegal hunting or opportunistic killing of lions, and the difficulties of regulating legal hunting, where it exists, at levels that can be considered sustainable and in ways that benefit local people rather than just international safari operators. Anger and emotion complicate efforts to keep lions alive and to find social consensus on the best ways of doing that.

The basic concept of Lion Guardians—local hunters, trackers, or warriors converting their skills and purposes to protecting wildlife from people and people from wildlife—is potent enough to take hold in other contexts. It's happening, for instance, at the Reteti Elephant Sanctuary in Samburu County of northern Kenya. No one is better suited to safeguarding large, inconvenient, dangerous animals, at least potentially, than people with centuries of tradition living in their presence.

But the population status of lions throughout Africa—in those five stronghold sites I mentioned (page 205) and in the 62 lesser enclaves of lions on the continent identified at the time this story first appeared—remains fraught. Their estimated numbers are smaller today, because either lion losses have increased, or counting methods have improved and become less influenced by delusive optimism, or a combination of both. Recently I spoke with Phil Henschel, Panthera's regional director for West and Central Africa, the scientist with whom I visited a lion-survey project in northern Benin 10 years earlier. Yes, the lion strongholds are still strong, Henschel told me, at least relative to other habitat patches; those five big sites contain the greatest share of African lions. But the latest population estimates, both site by site and in sum, are sobering. The Ruaha-Rungwa eco-

system of western Tanzania, for instance: A decade ago it was thought to contain about 2,000 lions, but the estimate now is less than a thousand. The new counting effort there was led by Paolo Strampelli, then a Ph.D. candidate within the Wildlife Conservation Research Unit (WildCRU) of the University of Oxford (and now Dr. Strampelli), using a version of capture-recapture methodology in which the "capture" is a high-definition photograph of each lion, made with xenon-flash camera traps that can reveal individual markings such as whisker spot patterns. This technique allows for counting, and estimates derived from counting, more accurate and precise than the old ways.

Recent estimates from the Selous ecosystem of southeastern Tanzania are even more distressing. Presumed previously to harbor between 5,000 and 7,000 lions, this "stronghold" is now thought to contain only about a thousand. Longa-Mavinga National Park in southeastern Angola, never a stronghold but reported by government sources to contain about 2,000 lions, is now reckoned to hold only about 20 or 30. The realistic total estimate of African lions in the wild, Henschel told me, has fallen from 35,000 to less than 20,000.

But there is also some good news, at least on the margins. In Zambia's Kafue National Park, that country's oldest and largest, lion numbers seem to be up. In Senegal's Niokolo-Koba National Park, the biggest such area in West Africa, where Panthera (with financial support from the Lion Recovery Fund) has played a role in training and equipping more rangers, the small and beleaguered lion population, counted by dozens, has more than doubled in a decade. In Gabon, where Henschel himself is based, Batéké Plateau National Park was one of those 13 parks created by Omar Bongo in 2002 (see "A Country to Discover," pages 81–96). A savanna ecosystem of grassland and woodland patches, streaked with gallery forest and containing a wide variety of wildlife, from duikers and jackals to elephants and chimpanzees, Batéké seemed to be empty of lions. "Then suddenly, in 2015, a lone male showed up in Batéké,"

Henschel told me. Presumably that lion had wandered in from the Republic of the Congo, just across Gabon's southeastern border. Male lions, especially young ones, sometimes disperse long distances in their search for territory, food, and females. The scientists weren't sure this one could find sufficient prey in Batéké. "But he has taken up residency." That opens the possibility of government approval for translocating females into the ecosystem, to meet the third of those three S's that Iain Douglas-Hamilton noted, as essential for lions as for elephants: security, sustenance, and sex. It had been done in Malawi, Henschel said, at a site where lions were now again resident and breeding. "We have the plan to do the same in Batéké."

This is lion conservation in the 21st century. Sometimes, in some places, you count them by the marks on their faces and your counts may build into the thousands. At other sites, even a single male lion turning up in suitable and protected habitat creates an opportunity that can't be ignored. No creature embodies wildness more vividly than the African lion, and every individual is precious.

ELEGY FOR
A LION

◆·◆·◆

The Death of C-Boy

The sad but inevitable news reached me through photo editor Kathy Moran, who got it by email from Daniel Rosengren, the field assistant (to lion biologist Craig Packer) with whom I had spent time in the Serengeti five years earlier: C-Boy, the noble protagonist of the story that Nick Nichols and I published back then, was dead. Daniel, having himself moved on to other adventures in other places, had heard it from a tour guide on the Serengeti. It was a consolation, as Daniel said, to know that C-Boy had died, naturally and with leonine dignity, rather than not to know and "keep wondering if he is out there." C-Boy had already exceeded all expectations.

Kathy connected me with the people handling online content for National Geographic, *and I wrote this short piece to go up promptly on the website. It ran accompanied by the magnificent black-and-white portrait of C-Boy, by Nick, that had run with the original Serengeti article. In it, as you'll see, I recapitulated a bit of what I'd written earlier, because those five years had passed and the audience was not identical. I ask you to take this, not as repetition, but as a sort of respectful funerary summation.*

◦◦◦◦◦◦◦◦◦◦◦◦◦◦◦◦◦◦◦◦

Deceased: Adult male lion, roughly 14 years old, with a dark mane, known to lion researchers who long followed his career—and to readers of *National Geographic* magazine—as C-Boy. Dead of natural causes, his body discovered by a tour driver in the backcountry of

Serengeti National Park, Tanzania, in early June 2018. His demise mourned by those who knew and read about him; his longevity and force of character a marvel to same.

The category "natural causes," in the case of African lions, includes the kind of murder and mayhem that occurs routinely among competing members of the species. As the lion expert Craig Packer once told me, "The number one cause of death for lions, in an undisturbed environment, is other lions." This was about five years ago, when photographer Nick Nichols and I were in Tanzania, doing fieldwork for a story on lion behavior and ecology. C-Boy, a handsome male in his prime, with a black-fringed mane, became the central figure of that story—titled "The Short Happy Life of a Serengeti Lion"—because he stood as an exception to this mortal rule.

Several years prior, C-Boy had barely survived a gang attack by three other males, who tried to kill him over mating rights to a pride of females. Those three ambitious males, along with one other, were known as the Killers. A field assistant on Packer's long-term study, Ingela Jansson, witnessed the three-on-one brawl from nearby in her Land Rover, saw C-Boy's wounds, and figured he was a goner. (It was Jansson who, after earlier observations of another trio, had "boringly" named them A-Boy, B-Boy, and C-Boy, not realizing then how extraordinary the third member of that alphabetical list would be.) But C-Boy slouched off the field of battle and, with his sole coalition partner, known as Hildur, wandered elsewhere to seek new territory, new females, and new prospects. That was nine years ago.

Myth holds that cats have nine lives. C-Boy had at least two. He endured the immediate attack, escaped a lingering death from infected wounds, and later became the starring character of our story. Why did Nick and I choose to focus on him? Because he was everything an African lion should be: resourceful, cantankerous, patient, proud but pragmatic, seemingly indestructible, continually imperiled, and gorgeous to behold.

During our fieldwork, the Killers turned up in an adjacent area, showing interest in another pride among whom C-Boy and Hildur had been fathering cubs. They were pushing again for new conquests, the Killers, threatening to expand their domain. Another assistant on the Packer study, a young Swede named Daniel Rosengren, spotted them at dawn one morning, as I rode along, where they lay on a grassy stream bank, nursing facial wounds from a recent fight. Whom had they fought? Our guess was C-Boy, again. Had he survived once more? If so, in what condition?

There were no answers through a long day of fruitless searching. Nick's team couldn't find him, and neither could we. Late that evening Rosengren and I equipped ourselves with night vision binoculars and sleeping bags, then rolled slowly overland behind the Killers in his Land Rover for the entire night, trading shifts of sleeping and watching, while the lions prowled, rested, and moved again. I called it the Night of the Long Follow.

These ambitious lions were on the march through C-Boy and Hildur's territory, and we wanted to see where they went, what they did, and whether their daring incursion—plus their battle wounds—meant that they had killed their way to preeminence hereabouts as well. Dawn came, the Killers walked boldly away down a dirt track, and for two days afterward, there was still no sign of C-Boy. In a journal entry, I recorded him as "missing, suspected dead."

But he wasn't dead. On the third morning, near a group of rock outcrops known as the Zebra Kopjes, we found him, unscathed and lusty, preparing to mount a ready female. In my journal entry for that day, December 17, 2012, I wrote: "O happy lion!" His mane showed dark and virile in the early morning light. He was very much alive.

But even C-Boy couldn't live forever. Last summer I got an email from Daniel Rosengren, now employed as a roving wildlife photographer by the Frankfurt Zoological Society. He confirmed what I had heard elsewhere. "Yes, C-Boy was found dead by a tour driver who knew him well," Rosengren wrote. "I can't really say much more than that. He had apparently already been dead for a couple of days when they found him (following the vultures that ate the carcass)." There was no sign he had been speared by a Maasai herdsman—intent on protecting cows—or shot by a poacher.

"He was about 14 years old," Rosengren told me, "touching the record age for a male lion in all the history of the lion project." Twelve years is generally the maximal lifetime for a male. C-Boy's partner Hildur, also pushing the limits, had somehow outlived him.

It was saddening, Rosengren said, to realize that C-Boy was gone. "But at the same time, he lived a longer life than expected for a male lion. A life that almost ended close to a decade ago when the Killers got him. He got a second chance and certainly made the best of it." Rosengren added: "I wish that I could have seen him one more time."

I wished the same, and knew I couldn't, so I did the next best thing. I opened the August 2013 issue of the magazine, and there was Nick's magnificent portrait of C-Boy, with his dark-fringed mane, staring back at me through the Tanzanian night. It consoled me with the reminder that C-Boy's life, short or long, happy or fraught, embodied a magisterial will to survive.

....

It's C-Boy, in a color version as photographed by Nick, who graces the cover of this book.

....

DESPERATE PRIMATES

-◆-◆-◆-

Chimps and People on a Ragged Landscape

. . . .

The conflict between wildlife and humans, which defines landscape battles all over the planet, is never more keen, raw, and discomfiting than when it involves our closest living relatives, the great apes: gorillas, bonobos, orangutans, and chimpanzees. Besides being highly intelligent, these creatures are startlingly strong, and willful, and can be gentle one minute, dangerous the next. Respectful appreciation of them demands remembering that, despite the high similarity of their DNA and ours, they are wild animals.

Maybe that should be a broader reminder: We humans are wild animals too, possessing technology and language and the ability to ratiocinate that only makes us more dangerous to other creatures and to ourselves.

Ronan Donovan, another of the superb National Geographic *photographers with whom I've been privileged to partner, brought this story idea to the editors and me. I knew Ronan from having worked with him on a special issue of the magazine devoted entirely to Yellowstone, during the research for which he and I jumped out of a helicopter together on a snowy mountain to follow biologists who were radio-collaring wolves. We also skied to a remote lake in the Yellowstone backcountry, again with biologists, to see a site where grizzly bears have learned to subsist, during the hungers of spring, by digging and eating earthworms.*

For this story, Ronan and I traveled to Uganda, where he had once worked as a young field assistant to a chimpanzee researcher.

There in western Uganda, chimps were now facing a very different sort of subsistence challenge.

· ·

Life was already hard enough for Ntegeka Semata and her family, scratching out a subsistence on their little patch of garden land along a ridgeline in western Uganda. They could barely grow food for themselves, and now a group of desperate, bold, crop-raiding chimpanzees threatened their livelihood, maybe even their safety.

The chimps had been coming closer for a year or two, prowling all throughout Kyamajaka village, searching for food, ripping bananas from the trees, grabbing mangoes and papayas and whatever else tempted them. They had helped themselves to jackfruit from a tree near the Semata house. But on July 20, 2014, scary tribulations gave way to horror—a form of horror that has struck other Ugandan families as well. That was the day when a single big chimp, probably an adult male, snatched the Semata family's toddler son, Mujuni, and killed him.

"A chimpanzee came in the garden as I was digging," Ntegeka Semata recalled during an interview in early 2017. Her four young children were with her that day, as she combined mothering with hard fieldwork, but she turned her back to get them some drinking water. The chimp saw his chance, grabbed her two-year-old son by the hand, and ran.

The boy's screaming brought other villagers, who helped the mother give chase. But the chimp was rough and strong, and the fatal damage occurred fast. "It broke off the arm, hurt him on the head, and opened the stomach and removed the kidneys," Semata said. Then, stashing the child's battered body under some grass, the chimp fled. Mujuni was rushed to a health center in a nearby town, Muhorro, but that little clinic couldn't treat an eviscerated child, and he died en route to a regional hospital.

Things were still uneasy in Kyamajaka when Ronan Donovan and I arrived, a few years later—uneasy for both people and chimpanzees. Attacks by chimps on human infants had continued, totaling at least three fatalities and half a dozen injuries or narrow escapes in greater Muhorro since 2014.

The main driver of the conflicts, it seemed, was habitat loss for chimps throughout areas of western Uganda, forested lands outside of national parks and reserves that have been converted to agriculture as the population has grown. The native forest that once covered these hillsides was largely gone, much of it cut during recent decades for timber and firewood and cleared to plant crops.

Such demographic and landscape changes are happening fast throughout Kagadi District (including Kyamajaka), which lies just east of Lake Albert and the Rwenzori Mountains, and in neighboring districts as well. The soil is volcanic and rich, well watered by seasonal rains, and suitable to support a burgeoning number of farming families that eke out their livings on small private plots from staple crops such as corn and cassava, supplemented by domesticated fruits and a little income from cash crops: tobacco, coffee, sugarcane, and rice.

The Uganda Wildlife Authority (UWA) is acutely aware of the chimp problem, and though chimps outside protected areas (as well as within national parks and reserves) fall under the authority's responsibility, private forests do not. "Unfortunately, it is hard for us—impossible for us—to prevent clearing of these areas," UWA Executive Director Sam Mwandha told me. "We can only plead. We can only educate and hope that people will appreciate them."

But appreciating a forest for its long-term benefits, such as mitigating erosion and buffering temperature, can be difficult in the face of short-term pressures to grow crops for food. And with chimps in a forest patch, one moment of diverted attention by a mother as she gardens can result in a child being snatched. So the immediate need, Mwandha said, is to "create awareness" among people in such areas that their caution must be high, their vigilance continuous. That's easier said than done, but the UWA has assigned four permanent staff to this awareness campaign in western Uganda.

The chimps of Kyamajaka—maybe just a dozen or so in the village environs—nest nightly in the remnant woods at the bottom of a glen, where a small stream runs, or in the eucalyptus plantation nearby. By day, they emerge, because their wild foods have largely disappeared, and feed from the crop fields and fruit trees surrounding village homes. (Imagine, in your own life, stepping out to weed the tomatoes and encountering a hungry cougar. Bear in mind that an adult male chimp can be more dangerous than a cougar.) They move stealthily amid the village, mostly on the ground because there's no forest canopy left for them to travel through, high and confident, as they would in deep forest. Despite the stealth, their pedestrian foraging sometimes brings them into close contact with people. They drink at the same stream where women and children fetch water. When they stand, or walk upright, as they often do, they seem menacingly humanoid.

Chimpanzees, along with bonobos, are our closest living relatives. Their species, *Pan troglodytes,* is classified endangered by the International Union for Conservation of Nature. Their total population throughout Africa—at most 300,000, possibly far less—is smaller than the human population of Wichita, Kansas. As adults, they are big, powerful animals—a male might weigh 130 pounds and be half again as strong as a similar-size man. Chimps in productive forests live mostly on wild fruit, such as figs, but they will kill and eat a monkey or a small antelope when they can, tearing the body to pieces and sharing it excitedly. They relish meat. Because chimps tend to be wary of adult humans, especially men, their aggressive (and in some cases predatory) behavior toward people, when it occurs, falls mainly upon children. In some cases, too, a chimp might pick up a small child out of sheer curiosity, as though grabbing a toy.

Whatever the motive, it can be terrifying. For more than three years after the trauma of her son's abduction, Ntegeka Semata and her husband, Omuhereza Semata, a farmer, continued to live in their house. They built a bamboo fence around their tiny backyard, enclosing the cooking shed in what they hoped would be a safe zone for the family. "I am scared all the time that other chimpanzees might come back," Ntegeka said in that earlier interview.

But the fence was flimsy, the chimps kept returning, and the Sematas felt under siege. Ntegeka couldn't work in the garden. The children were sometimes too afraid to eat. Even their goat made piteous noises of fear. By the end of 2017, their house was vacant, with a broken window above the front door. The Sematas had fled and were living a marginalized existence in a rented room at a compound three miles away. They owned no farming land there. "I feel like we've been cast back into poverty," she said.

Meanwhile, the remaining windows of their old house reflected only the faces of chimpanzees, which visited regularly, glowering in, confused and provoked by the chimp images mirrored there, which seemed to be glowering out.

THE DEATH OF Mujuni Semata was no isolated event. Police reports from the town of Muhorro (of which Kyamajaka is a satellite village, containing a few hundred families) described two chimp-on-child attacks during 2017. On May 18, a toddler named Maculate Rukundo was seized in a cornfield while her mother worked the crop. The mother chased the chimps but then backed off, terrified, and ran to get help. A crowd of local people, soon joined by police, tracked the chimps to a patch of forest, where the little girl lay dead in a pool of blood and intestines, her gut torn open by chimp fingernails. Five weeks later, chimps (maybe the same group, but that's hard to know) took a one-year-old boy from another garden plot, with his mother nearby, and again retreated to a patch of forest. A posse of local villagers pursued the chimps until they dropped the boy, who had a deep cut on his left leg but was alive. The police reported that, in addition to this survivor with serious injuries, six young children had been killed in the area by chimps.

Then, in mid-2018, a five-month-old girl was snatched from a veranda while her mother worked in the kitchen. The mother heard her child's cries, raised a ruckus, and charged the chimps—and they fled. That baby was found alive, unconscious, in a nearby bush. After going to a well, a three-year-old girl was taken by a male chimp that scared away the child's older friends and carried her off but dropped her, reportedly when he was challenged by an elderly man, a passerby, who raised the alarm. A 12-year-old boy in another satellite village was grabbed near a garden and suffered a deep arm wound as he struggled to get free.

From elsewhere in western Uganda have come accounts of the same gruesome pattern, played out with variations: one child killed by a chimp on the sugarcane plantation at Kasongoire, in 2005; four chimpanzee attacks on children, with one fatality, near the Budongo Forest Reserve, farther north; eight attacks, back in the 1990s, seven of which were probably by a single rogue male chimp, on children from villages bordering Kibale National Park. Of those victims, three children were eviscerated, and some were partly eaten. That male, further demonized with the name Saddam, was hunted down and killed soon after his seventh child killing. He was an egregious anomaly. Most cases are more ambiguous, involving chimps that are reckless at one fateful moment, not repeat killers. And this phenomenon isn't confined to Uganda: It has happened elsewhere in chimp range across Africa, most notoriously at Gombe Stream National Park, Jane Goodall's study site in Tanzania, where in 2002 an adult male chimp snatched and killed a human baby.

Chimpanzees in Uganda are protected by law, meaning that it's illegal to hunt or kill one, regardless of whether it lives within a park or reserve (though permission has occasionally been granted to kill a rogue male such as Saddam). They are further protected by tradition of the Bunyoro people, predominant in western Uganda, who tend to see chimps as different from other animals and, unlike some Congolese peoples across the border, don't hunt them as food.

Despite law and custom, in this part of Uganda, there have been killings of chimps too—retaliatory, defensive. The details will probably never be known. In December 2018, an adult male chimp was fatally speared. A young female was beaten to death with sticks and stones. The carcass of another young chimp was reportedly found, decomposing, cause of death indeterminable but fingers cut off. Among communities of angry, powerless people who fear for their children, it's not surprising. Chimpanzees aren't the only desperate primates in western Uganda.

All these painful ambiguities show up vividly at a place called Bulindi, where one group of chimpanzees and their fraught interactions with people are studied by a British biologist named Matt McLennan.

McLennan came to Uganda in 2006, as a doctoral student at Oxford Brookes University, in England, to study how chimpanzees adapt their behavior to living in a human-modified landscape. Why that topic? Because he foresaw the challenges to come for chimpanzees everywhere. He knew that the Budongo Forest Reserve was good habitat containing about 600 chimps and that another forest reserve 50 miles to the southwest, Bugoma, harbored roughly the same number. Between those two refuges, Budongo and Bugoma, was a mixed landscape of small farms and large sugarcane plantations, with a growing human population and shrinking strips and patches of forest, which had once represented a connecting zone for the two reserve-protected populations and latterly sheltered small resident groups mostly isolated in remnant patches of habitat. A total of about 300 chimps lived within that middle zone, finding refuge in the forest patches, venturing out onto croplands for food. Some individual chimps—young females, for instance, escaping their fathers and brothers to find new mating possibilities—would move from one small group to another, or even from an isolated group into Budongo or Bugoma, providing some gene flow; but as the forest patches shrank and isolation increased, even that modest degree of intermixing became difficult.

Much of the land was private, loosely held by customary occupancy and inherited through the male line. After passage of the 1998 Land Act, which formalized traditional tenure in Uganda into deeded property, people felt greater security of ownership. That security, ironically, empowered them to harvest their forests and switch to crops. Survival amid such a landscape, for a single chimp or a group of them, was problematic.

This tangle of circumstances drew Matt McLennan to Bulindi, a town on the road about halfway between Budongo and Bugoma, where he found a group of at least 25 chimps. With a local research collaborator named Tom Sabiiti, he began work, the first step being to persuade those animals to tolerate his and Sabiiti's presence in the forest. He wasn't trying to habituate them and make behavioral observations; instead, he wanted to gather ecological data from indirect evidence such as fecal samples and nest surveys. Still, it was difficult. Unlike wild chimps in good, expansive habitat, which tend to be shy, these Bulindi chimps had a belligerent edge.

"We found out pretty quickly that they didn't like people inside the forest," McLennan told me. "Their strategy was to try to intimidate us. Which they did very effectively." The big males especially: They hooted, drummed on the ground, thrashed vegetation. One day they chased McLennan 250 yards but left him unhurt when he fell. Eventually the chimps grew sufficiently inured to his and Sabiiti's presence that they tolerated it without responding aggressively, and the pair gathered data for two years. At that time, McLennan recalled, a fair bit of forest still spanned the hillsides and shaded the stream valley draining through Bulindi, though clearance was under way, and the sound of chain saws rang out in the woods. Farming was mainly for subsistence, but cash crops (notably coffee and tobacco) had arrived. And the chimps were getting bolder. The first chimp attack on a child, within memory of local people, occurred in 2007. The following year, McLennan went back to England and wrote his dissertation. When he returned in 2012 to continue field research on the Bulindi chimps, things had changed.

Most of the forest was gone. Crop fields now spread widely across the hillsides above the small stream: corn, cassava, sweet potato, and other garden produce. There were fewer chimps in the local group and, among those still there, fewer adult males. Some of that decline may have been deaths from leghold traps, an illegal and sometimes lethal means of discouraging crop raiders such as chimps and baboons. The remaining chimps seemed even bolder than before, at least around women and children, but their boldness was somewhat less aggressive. Their diet included more of the human crops. They had begun eating jackfruit, a new behavior since 2006, and local residents resented their jackfruit losses. McLennan decided that, rather than bemoaning these changes, he would study how the chimps were adapting.

What he found was that the chimps at Bulindi were coping, at least temporarily. Their number was up slightly, from 19 in 2012 to 21 by the end of the decade. Their condition was generally good: They were robust and strong. Most adult females had infants. Genetic analysis of the chimps' DNA, from fecal samples, suggested that their isolation hadn't yet brought severe inbreeding—although, according to Maureen McCarthy at the Max

Planck Institute for Evolutionary Anthropology, in Leipzig, Germany, who led the genetic study, that could change with increasing isolation, decreasing female dispersal, and time.

But the Bulindi chimps did carry higher levels of stress-related hormones, at least during some times of year, than a population of chimps within the Budongo reserve, just 20 miles away. Did that mean their piratical way of life, staying so close to humans and raiding their food, was inherently stressful? Maybe, although other complex variables also affected those hormone levels. It was hard to know whether the Bulindi chimps were thriving on human foods, suffering tension from their nearness to humans, or both.

Among the people at Bulindi, attitudes toward the chimpanzees varied. One woman told me she wished they would stay in the forest. Her husband interjected: "The forest is over." Another woman considered them a small nuisance for stealing her jackfruit and bananas, but at least they kept the baboons away. An amiable matriarch named Lillian Tinkasiimire, whose little redbrick house was graced with a mango tree in front, a fig tree behind, both of which attracted chimpanzees, took a steady view.

"The chimps are very clever," she told me. "If you don't chase them, they will be your friend. If you chase them, you will see fire." Tinkasiimire had preserved much of her forest. Her attitude was, Let the chimps live there, let them be, let them visit.

McLennan hoped to encourage such tolerance and help make it less costly. He and his fiancée, Jackie Rohen—a writer trained in musical theater but now committed to the theater of conservation—also created the Bulindi Chimpanzee and Community Project. It provides development assistance to families in the area and incentives to mitigate human-chimp conflict: payment of school fees in exchange for reforestation, for example, and starter plants for shade-grown coffee, fuel-efficient stoves that use less firewood, new borehole wells (as an alternative to stream pools) that allow women and children to avoid chimpanzees when fetching water. The best way to preserve peace between Bulindi's people and its chimps, McLennan and Rohen recognized, was to help them stay apart.

At Kyamajaka and other villages near the town of Muhorro, three hours southwest of Bulindi, things were different. Matt McLennan didn't study these chimps, and no similar community project offered incentives to preserve forest or measures to defuse conflict. No one knew how many chimpanzees lurked or cowered in the Muhorro forest remnants (maybe 20, maybe fewer?) or where their next unfortunate conflict with humans might occur. The Semata house stood vacant and solitary after the family's departure, with chimps coming there frequently, more than a dozen individuals—as documented by Ronan Donovan's photographs—to menace themselves in the reflective windows and kick their feet against the walls. What person would want to live in such a place?

Across the glen, half an hour's walk down one garden hillside and up another, Ronan and I spoke with a man named Swaliki Kahwa, whose son Twesigeomu (known as Ali) was taken by a chimp a year earlier, before his second birthday—dragged away and fatally battered. Kahwa deferred to his elder brother, Sebowa Kesi Baguma, the village chairman, to tell us about it. Baguma, a grave but cordial man wearing a yellow T-shirt and green gum boots, produced a police report and showed us the postmortem photos, printed in shadowy but lurid magenta. The boy's right arm had been nearly torn off; a gash on his right leg, near the groin, may have cut the femoral artery; some of his fingers were broken. According to the report, which included timing of events, little Ali took almost 12 hours to die.

Baguma noted drily that people of his village have been taught to consider chimpanzees "beneficial." This was the message from one international conservation group with activists in the area and from others who imagine chimp-based ecotourism bringing visitors to the cornfields around Muhorro. "We don't see any benefit," said Baguma. "It's killing our children."

The national reserves, such as Budongo and others, with sizable chimpanzee populations, are problematic for the Uganda Wildlife Authority. Those areas are degraded by illegal woodcutting, cropping, and settlement, with which the agency, in partnership with the National Forestry Authority, deals firmly. Some illicit settlers are even evicted from the reserves. But for chimp-human conflict within communities such as Kyamajaka, garnished with scraps of private forest, UWA's approach is gentler, as described by Executive Director Mwandha: creating "awareness" of the immediate dangers and potential benefits of chimpanzees amid villages, and patrolling to monitor chimpanzee presence.

Whether such awareness can change attitudes in the more traumatized communities, with children and chimps still in harm's way, is an urgent question. Back across the glen, after listening to Baguma's anger, Ronan and I encountered Norah Nakanwagi, the chairwoman of Kyamajaka, as she sat outside her house, resplendent in a black bandana and a floral blue outfit with puffy shoulders, the sort of formal dress known as *gomesi* in Uganda. She spoke in Runyoro, the Bunyoro language, through our translator. It's unsafe here for women and children, she said. She waved her hand at a cornfield. I can't go there. We've had five children killed since 2007, she said. People tell us the chimps are beneficial. Yes, we should leave them alone, but it's difficult to explain that to someone whose child is dead.

Then she switched to English: "Take them away. Not to kill them. But take them away."

Why not move the chimps? People do ask about that, McLennan told me. But move them where? There's no vacant chimpanzee habitat anywhere in Uganda. And dropping them into occupied habitat would be murderously stupid, provoking chimpanzee war. Another dire option: Kill the chimps, fast and cleanly, to protect the people and put the chimps out of their misery. But are they in misery, with their high body fat and healthy reproduction, fueled by pilfered mangoes and jackfruit?

No one is likely to advocate killing these chimps, dangerous though they may be, as official Uganda policy. Once adopted, where would that line end? Anyway, there's a third option: trying to manage the situation somehow. Small projects, reforestation incentives, tactical mitigations, borehole wells, alternate sources of income, patience, sympathy. Creating greater awareness, as the Uganda Wildlife Authority suggests, of the immediate dangers and how to avert them, as well as the long-term possibilities, if any, of economic benefit from small-scale tourism. Incremental but tireless efforts to help chimps and humans observe an uneasy truce.

It's a local problem that's not just local. Uganda's awful dilemma fore-tells the future of chimpanzees across Africa. Less forest, more people, more desperation among the chimps, more conflict. What makes a village like Kyamajaka seem so pitiable, and a town like Bulindi seem so important, is that in those two places the future has arrived.

....

The first point to be noted as coda to this story is a sad one: Jackie Rohen, the partner of biologist Matt McLennan and the "driving force" behind the Bulindi Chimpanzee and Community Project (to quote McLennan), died suddenly of a pulmonary embolism before this story reached print in National Geographic. *Jackie was charming, smart, lively, dedicated, and had literary interests and talents as well as a deep commitment to the chimpanzee work—her first novel, titled* How to Marry Your Husband, *was in the pipeline at Penguin Random House, to be published just three days after what turned out to be the day she died. She gave a large part of her heart and attention to the Bulindi communities, both human and chimpanzee, even while she was busy with further literary ideas; and she offered generous help and hospitality to Ronan and me while we researched and photographed this story. She is remembered with admiration, and missed, on at least three continents.*

I heard more from Matt McLennan recently, about both her and the project. Jackie was "full of ideas and practical skills," he wrote, "and she put all the organizational structures in place. We all learned so much from her, which is how BCCP was able not only to stay afloat but to continue in the direction she had planned for it." This is an important addendum, reminding us that being a field ecologist is a demanding job—long hours of tracking animals and inferring their activities, recording and analyzing data, writing journal papers and grant applications—and that to launch and maintain an addi-tional effort devoted to the human community in the ecological context, an enterprise such as the Bulindi Chimpanzee and Community Project, is more than work for a single person. It wouldn't have happened without Jackie Rohen.

But it has happened, and it continues to achieve results. As of 2022, McLennan told me, the BCCP suite of programs has expanded to include health and behavioral monitoring of five groups of chimpanzees, a tree-planting initiative that has put four million trees into the ground, more school-fees sponsorship of community children, construction of 40 borehole wells to defuse the conflict with chimps over stream water, village loan groups supporting environmentally friendly household enterprises, a school outreach program teaching child safety amid chimps, and a football league, named in honor of the chimps and comprising 18 teams, sponsored with jerseys and balls and tournaments, to brighten the emotional valence of the "other" primate on the landscape. McLennan added that the "forest in Bulindi is regenerating and looks in better condition than it has done in years."

But difficulties remain, one of which is that the demographics of some chimp groups have changed, with time, in the context of increased proximity to humans on the changing landscape. In the local Bulindi group, for instance, the old alpha male, known to McLennan and his team as Sylvester, was deposed and disappeared. In his place rose an excitable young male, known as Moses. In addition, the group includes three other young adult males and only one seasoned codger. "These young males," McLennan told me, "have grown up in a village environment; they've never known anything else and they aren't afraid of most people. Moreover, young male chimps are bursting with testosterone and keen to impress group mates. It's fair to say these young males are causing havoc, roaming around homes with impunity." Since 2019, two or three children each year—just in the Bulindi area—have been injured in chimpanzee snatch-and-drag encounters. And in May 2022, a fatality, the first Bulindi child: a seven-month-old baby.

Beyond the Bulindi area, in a different village and involving another chimp group that McLennan's team now monitors, a one-year-old baby was killed the year before. Local people responded to that event by hunting and killing an adult female chimp, known to McLennan as Maggie. Maggie's own five-year-old orphaned child is now seen following the older males. It's a cycle, like other cycles of violence, in which nobody wins. But community projects that deliver both concrete benefits and interspecies sensitization can mitigate the strife.

"For now, the fragile coexistence between people and the village chimps looks set to continue," McLennan told me, "and we'll continue to play our part—trying to make the situation better for people and chimps—as long as we're needed."

THE MEANING OF NORTH

◦|◦|◦|◦

Pristine Seas and Lonely Islands

. . . .

The terms of this assignment gave me pause: You will be on a boat in the Russian Arctic, with an international team of scientists, most of whom will scuba dive in the frigid waters to assess biological diversity, and you will be out for six weeks. The boat leaves from Murmansk, northwestern Russia's major port city. Six weeks sounded longish, given that I am very happily married, to a formidable and adventurous soul named Betsy Gaines Quammen, and for marital accord I try to avoid work trips lasting more than a month. I described this expedition to Betsy and asked her response. I think you should do it, she said. Furthermore, she added, if they had invited me, I would already have said yes.

My photographic partner for this caper was Cory Richards, a wizard expedition photographer and a Himalayan climber, roguish and charming. We were nearly strangers when the trip started and good friends at its end. There was other fine company, among the scientists on board, including one unexpected addition: Mike Fay, invited along to help with the botanical surveying on land. I always enjoyed being in the field with Fay—a man who combines deep ecological knowledge with quiet physical toughness and an all-in commitment to conservation—but he was especially welcome to me as company on this trip because he had no more intention of scuba diving in Arctic waters than I did. We would walk the shorelines and basaltic plateaus of these austere islands, with a few similarly inclined Russian colleagues, while most of the others swam.

. . . .

Feodor Romanenko, wearing his usual puckish smile, the one that looked like it could have been carved with a melon scoop, raised his arms. "Dear colleagues," he said, and then launched into his Russian-accented French. "Dear colleagues" weren't quite the only words of English he knew, but they were clearly his favorites, useful for summoning attention from a motley international group such as ours. *Dear colleagues, I propose that we now climb up there,* he would say, indicating a precipitous, unstable, ugly hillside of scree. *Dear colleagues, lunchtime! Let us enjoy it here atop the butte before high winds and the next snowstorm arrive. Dear colleagues,* he bragged cheerily to our evening assembly, *today my group made five wondrous discoveries, including two kinds of basalt! and some Mesozoic sediments! and evidence of recent deglaciation!* Romanenko, a geomorphologist based at Moscow State University, had 28 seasons of experience (at that time) on the shores and the islands of the Arctic Ocean, and his enthusiasm for his work was still boyish. Trudging across a severe northern landscape, he exuded contagious joy in doing field science—making close observations, seeing patterns, and compiling data that could help answer, among other mysteries, the question of ice.

Contagious joy is a valuable commodity in the Russian Arctic, more precious even than vodka or chocolate. We had come north with Romanenko, a day's sailing out of Murmansk, to an archipelago known as Franz Josef Land, and although it was not our primary purpose, that question of ice underlay much of what we were here to learn. It was really three questions: Why is the perennial ice melting? How far will that melting go? And with what ecological consequences? When you make a biological expedition into the high polar regions, Arctic or Antarctic, in this era of climate change, the question of ice is always important, whether or not you address it directly.

Our approach was indirect. We were 800 miles north of the Arctic Circle, almost 40 of us, members of the 2013 Pristine Seas Expedition to Franz Josef Land. Our mission was to view this remote archipelago through a variety of lenses—botanical, microbiological, ichthyological, ornithological, the plastic window of a scuba mask, and more. Franz Josef Land comprises 192 islands, most of them built of Mesozoic sediments covered with a capping of columnar basalt, and so flat across the top that, viewed without ice (as they increasingly are), they look like mesas or buttes in Arizona. Throughout earlier times they supported no permanent human habitation—until the Soviets established research stations and military bases on a few of the islands. That presence shrank during the 1990s, when the Russian government shrank, but increased thawing, new sea routes, and economic considerations were bringing renewed attention to this area by Moscow.

For more than a month, we zigzagged through the archipelago, drawn here and there by opportunity, driven by weather, escaping the winds that pushed the brash ice (floating fragments) and the bergs, going ashore when the polar bears let us, admiring the walruses and ivory gulls and bowhead whales, gathering data in places where few data have ever been gathered.

Our ship was the *Polaris,* a refitted tourist vessel with closets converted to laboratories, microscopes on dining tables, and an entire salon filled with scuba gear, including dry suits to protect our divers from water at 30°F. The team included Russians, Americans, Spaniards, Britons, one Australian, and a couple of Frenchmen. Each day some of us went ashore, on the latest island near which we had anchored, to walk transects, band birds, count walruses, or collect plants, while others dove the cold water to take inventory of marine microbes, algae, invertebrates, and fish. The walking days were sometimes long, but we were always back at the ship before dark, because dark never came. The sun didn't set; it just looped around irresolutely in the northern sky. The dives were short but dauntingly cold—I know this from testimony, not experience—even for a man wearing Ninja Turtles underwear beneath his dry suit. Feodor Romanenko's perspective on our meanderings was especially important, not just for science but also for morale, because it combined geology with élan.

Romanenko was not so space-age in style as the divers. In his floppy-eared hat, his iridescent orange vest, and his hip waders, with his shotgun in hand, he looked like an affable duck hunter from a small town in Minnesota. His other key piece of equipment was a garden spade. Katerina Garankina, one of his Ph.D. students from Moscow State University, red-haired and field-hardy, assisted him in the work of drawing geomorphological profiles of the islands. J. Michael Fay, doing the botany, was a natural on the daily outings ashore because, like Romanenko, he suffered an unquenchable craving to walk. Fay's epic survey hike across the forests of Central Africa ("The Megatransect" series, pages 19–80) was neither the first of his wilderness treks nor the last, and now that he was 58 years old, dividing his time between a cabin in Alaska and a conservation job with the government of Gabon, he was no less restless and impatient for foot travel through wild places. Arctic flora were mostly new to Fay, but on our first afternoon ashore in Franz Josef Land, I watched him identify a dozen flowering plants to at least their genera, each plant just a delicate clump of leaves within the pavement of rocks and mosses, its stems topped by tiny yellow or red flowers.

Now, nine days later, on an island called Payer, Fay was down on his hands and knees again, squinting, counting petals and carpels, taking photos. He had 12 species in his notebook by the time Romanenko and Garankina were done measuring the old marine terraces sloping up from the beach.

There were old marine terraces on Payer and elsewhere because Franz Josef Land as a whole has risen relative to sea level. It experienced episodes of uplift during the late Pleistocene and recent millennia, totaling, in some parts of the archipelago, more than 300 feet of elevation. The islands, on the far northerly wedge of the Eurasian plate, now ride higher in the water. Those uplifts have been driven by tectonic forces and to some degree allowed by the disappearance of ice. As a glacier melts away, its mass vanishing, its weight dropping, the terrain beneath tends to rebound, like the dent in a sofa cushion after you've gotten up. So the very shape of the landscape, not to mention the shape of the ecosystem it supports, is determined in part by the presence or absence of ice.

Ashore on Payer, I had been engrossed with Fay's identification of flowers, and scribbling notes of my own, until I heard Romanenko call our notice to a polar bear, huge and handsome, silhouetted on a ridgeline to the west. The bear seemed oblivious to us, but we knew better than to assume. As it walked, its small head surged forward on the rippling muscles of its long neck, suggesting the short-range striking speed of a snake. Our assigned guard and protector, a young man named Denis Mennikov, carried a Saiga-12 automatic shotgun with a banana clip, but the last thing we wanted was to bring that into action. Disappearing ice is a hardship for bears too, one that may force some reckless behavior. Dear colleagues, please be alert.

THE DYNAMIC VARIABILITY of ice is just part of what once made the Arctic, and Franz Josef Land in particular, so difficult yet enticing to explore. Fridtjof Nansen was only the most famous of the many explorers to touch the archipelago during some bold, miserable polar expedition. Things had gotten easier, not to say easy, since Nansen's desperate bivouac up here through the winter of 1895–96. For the Pristine Seas voyage we had better maps, lesser ambitions, GPS capacity, and a more comfortable boat. We also had a leader blessed with more aplomb than some of the bullheaded chieftains of the old efforts: National Geographic Explorer in Residence Enric Sala, a smart marine ecologist who pulled together this complex international effort, with support from the Society and other sponsors, as the latest in his series of Pristine Seas Expeditions.

Not many years earlier, Sala was a professor at the Scripps Institution of Oceanography, teaching grad students about food webs and marine conservation but dissatisfied with his contribution to the world. "I saw myself as refining the obituary of nature, with increased precision," he told me during a conversation aboard the *Polaris*. His distress at the continuing trends of ecosystem degradation and species loss, in marine as well as ter-restrial realms, led him out of academia. "I wanted to try to fix the problem," he said. So in 2005 he assembled a SWAT team of scientists, including experts on marine microbes, algae, invertebrates, and fish, and sailed for the northern Line Islands, a remote cluster of coral outcrops in the Pacific about a thousand nautical miles south of Hawaii.

There they dived and studied the reefs, making at least one important discovery: that predators, notably sharks, accounted for roughly 85 percent of the local biomass. That was topsy-turvy: Conventional ecological wis-dom posits roughly a 10-to-1 ratio of prey to predators at each level of a food web from bottom to top. Sala's team accordingly called this the inverted biomass pyramid. In the apparent absence of masses of prey, what could possibly sustain those abundant sharks? The answer was that the prey masses weren't really absent; they were produced copiously and continuously, in the form of small fish with high rates of reproduction, growth, sexual maturation, and turnover, and the predators continually cropped them to the point where they were inconspicuous. This is what ecologists call top-down regulation. It's a crucial thing to know about an ecosystem. Four years later, when outgoing President George W. Bush signed a bill establishing the U.S. Pacific Remote Islands Marine National Monument, Sala was in the room, and a mandate for preserving the inverted biomass pyramid was in the law.

With continuing support from the National Geographic Society, Sala took his Pristine Seas model to a series of other remote oceanic ecosystems, all in the tropics, where the waters were warm, fecund, rich with diversity, and clear. Then he turned his attention to the northernmost archipelago in the world, Franz Josef Land.

Franz Josef Land is a *zakaznik,* a nature reserve, administered within Russian Arctic National Park (though not officially included within the park), so Sala established a partnership with park officials and with the Russian Geographical Society. He enlisted Maria Gavrilo, an Arctic seabird biologist who served as the park's deputy director for science, to be co-leader. He rounded up some of the same intrepid researchers (including viral ecologist Forest Rohwer, fisheries ecologist Alan Friedlander, algae expert Kike Ballesteros, and Mike Fay) and trusted dive pros from earlier expeditions, and he welcomed a dozen Russian colleagues besides Gavrilo. He brought in Paul Rose, from the Royal Geographical Society in London, for his polar diving and climbing experience, his expedition-management expertise, his problem-solving skills, and his ineradicable good cheer. To this distinguished group, he added a handful of us media types. In late July 2013, we all sailed for Franz Josef Land, where the waters are certainly not warm or clear, and where the sea remained nearly pristine because for so much of the year, at least until recent years, it had remained largely covered with ice.

OUR TWO FRENCHMEN, David Grémillet and Jérôme Fort, had come to study the little auk *(Alle alle)*, a black-and-white bird that nests on cliffs and amid scree boulders and dives for its food in the frigid water. The little auk is still abundant throughout the Arctic, with a population estimated at more than 40 million—one of the most numerous seabirds in the world. But its family kinship with the great auk, an icon of human-caused extinctions—the last known pair was killed for a bird collector in 1844 off the coast of Iceland—serves as a reminder that no species is invulnerable to the pressures we humans generate. Beyond that, Grémillet and Fort had other grounds for focusing on the little auk. It's a tiny bird, as seabirds go, second tiniest of the auk family, with small wings that allow it to swim underwater as well as through the air. Its energy costs and its metabolic rate are high. So if its environment changes, Grémillet told me, the little auk may be more severely affected than other species. And its environment is changing—recent average temperatures in the Arctic are higher than they have been for the past 2,000 years. One study of Arctic trends has projected further increases of as much as 14°F by the end of this century.

The little auk feeds primarily on copepods, minuscule crustaceans that are the main component of Arctic zooplankton. Each bird needs to gobble thousands of them to make a square meal. "And these copepods, they have very specific temperature preferences," Grémillet explained. "So you can predict that if these copepod communities change because of climate change in the Arctic, the little auks will show a strong response."

How might the copepod fauna change? One of the larger and fatter kinds, a little thing about the size of a caraway seed, of the species *Calanus glacialis,* depends upon very cold water and the presence of sea ice, beneath which grow the algae that it eats. A smaller and leaner species, *Calanus finmarchicus,* is common in the North Atlantic and often rides currents into the Arctic but doesn't flourish there. As the Arctic Ocean warms by a few degrees, though, the competitive balance could shift. Higher temperatures and decreases in sea ice could allow the small, lean copepods to replace the relatively big, fat ones, to the detriment of the little auk—and of other creatures as well. Arctic cod, herring, and various seabirds feed on the copepods, and even such mammals as ringed seals and beluga whales depend on fish that feed on them. That's why scientists consider *Calanus glacialis* a keystone species in the Arctic.

Grémillet and Fort caught little auks by laying out patches of "noose carpet" in which the birds got their feet tangled, and then each bird was weighed, measured, and banded. Some birds were also fitted with a time/depth recorder or a geolocator, miniaturized units affixed to a leg or to breast feathers, from which data could be retrieved. The geolocators would track migration routes south after the birds bred. The time/depth recorders would reveal how deep a bird dived, how long it stayed down on each dive, and how many hours daily it devoted to such laborious food getting. From earlier work on Greenland and Spitsbergen, Grémillet and Fort knew that little auks with only *Calanus finmarchicus* (the lean copepod, not the nice fatty one) to eat must forage up to 10 hours daily to meet their energy needs. And those data were gathered in winter, when the birds had just themselves to support. How much worse might it be if, in summer, with chicks to feed and incubate, they had only that meager and labor-intensive source of food?

So far, little auks had shown admirable flexibility in the face of incremental change. But the question, Fort said, was: How much farther could they flex? "We think there will be a breaking point."

On a Monday in late August, after two tries, we succeeded in reaching Cape Fligely, on the north coast of Rudolf Island, the most northern of the group. Here, while the others were variously focused, Paul Rose and I escaped ashore for a hike to the top of the glacier.

We climbed up from the beach cautiously, because two polar bears had shown themselves hereabouts the previous night and one again this morning. But those animals seemed to have ambled away, and the coast was clear. As always, we had a security man: another young Russian, Alexey Kabanihin, who carried flares, a radio, and a Saiga-12, its clip loaded with blank rounds preceding the real ones. It was a glorious sunny day. From the western cape where we landed, a great dome of ice rose gently inland

and upward, a smooth arc sweeping toward nothingness like the curvature of the moon. Far below, afloat on the steel blue water, was the *Polaris*. In crampons, with ice axes, Rose and I started crunching up the slope, Kabanihin lagging behind. The ice was soft on its surface, beaded like corn snow, and sturdy beneath; the footing was good. After a day of shipboard confinement yesterday, Rose and I were thrilled with this getaway and could hardly control our foolish grins. But as we neared the top, a voice from Kabanihin's radio broke our mood. It was Maria Gavrilo, saying, "Paul, the polar bear smells you. And is coming toward you. Climbing the glacier. I suggest you come down."

We looked at each other. "Roger, Maria," Rose said. "That is all understood." He shut her off. We had no idea that she was coping with an ugly situation below—too many of us now on the island, spread out, unresponsive to cautions, and bears moving about. Could we go ahead just a little? Rose asked Kabanihin, who shook his head and gestured with crossed arms: absolutely *nyet*. But we were thinking: *da*. "One minute?" pleaded Rose. When the poor young man cringed indecisively, we both took off running. With a combined age of 126 but adolescents at heart, Rose and I galloped away, onward, beyond reach of authority and good sense, to a point very near—if not exactly—the highest spot on the northernmost landmass of Eurasia. We had summited. Get a GPS reading, I said.

He reported: 81°, 50.428 minutes north. Elevation: 174 meters. I scribbled those numbers in my notebook. Data. Then we ran back to Kabanihin, who looked unhappy, though not as unhappy as he soon would.

Descending over the curvature of the dome, we saw one polar bear between us and the ship, another bear off to our left. The bear in front was climbing toward us. The other was seated but cranking its head around slowly, like the turret on a tank, to follow us as we moved. I realized the situation was serious when Kabanihin handed me a flare. We shuffled on. Stay quiet, Kabanihin signaled. Stay close. He seemed very nervous. The glacier was big and open, and it belonged to the bears. We tried to angle between them, but the one ahead closed that angle, striding toward us with purpose. Suddenly I felt as if we were just three pieces of dark meat on a very white plate. I kept an eye on the left bear, expecting it to charge while Kabanihin was distracted by the other.

Kabanihin set his gun on the ice. He took back my flare, unscrewed the cap, and fired it toward—but not precisely *at*—the bear ahead. A red phosphorus Tinker Bell skittered out across the ice. When the bear scampered leftward a few dozen strides, we had an opening to go.

We were lucky. Getting ourselves killed, or getting a bear killed, Sala reminded us later, would have ruined the expedition. Even my adventurous wife, when she heard this story upon my return, thought we'd been inexcusably stupid, like the tourist in Yellowstone who tries to take a selfie with a grizzly. I had to admit she was right.

ON THE NORTHEAST coast of Heiss Island, near the middle of the archipelago, stood the remains of an old meteorological outpost known as Krenkel Station, which had pulsed with activity during the Soviet era. Established in 1957, it grew to include several tall antennas held up by guy wires, a launchpad for smallish research rockets, a miniature rail line for moving supplies and equipment, and dozens of buildings. At its peak, 200 people worked and lived at Krenkel. Now there were just half a dozen, and at least two dogs, a black-faced husky and a creamy one, which greeted us curiously at the beach when Romanenko, Garankina, Fay, and I jumped ashore.

Our presence had been cleared with the head of the station, and he left us to wander unsupervised through his little fiefdom of wreckage. Only the dogs came along.

The station thrived from about 1967 to 1987, according to Romanenko. Elsewhere in Franz Josef Land, a Soviet air base known as Nagurskoye supported long-range nuke-capable bombers that took off and prowled the Arctic in nervy readiness, during periods of high Cold War tension, just as bombers from American bases did. But Krenkel Station was not part of that. It was scientific in purpose and even modestly internationalist, through a collaborative arrangement with French meteorologists launching similar research rockets elsewhere. Then came the big changes toward the turn of the 1990s, as the Soviet Union approached its breaking point.

We can scarcely imagine, we who didn't go through it, what that final period of the U.S.S.R. was like: a stressful, confused, and anxious time (as well as a thrilling one, with its promise and possibilities) for many Soviet citizens, and no doubt especially hard in the boonies, as the distant central government metamorphosed so shockingly. (This effect on remote Soviet communities was what I saw also at Oktyabrsky on the Kamchatka Peninsula, described in "The Long Way Home," pages 147–164). Franz Josef Land is as far into the boonies as you can get. Making matters worse, in 2001, a fire devastated Krenkel Station. Personnel were pulled out and not replaced. They left their little houses, their recreation center with its two pianos and pool table and library, and they boarded boats or helicopters that carried them back to the mainland. Romanenko seemed to see all that in his mind's eye as we walked amid the wrecked trucks, abandoned tractors, pulled engines, spools of cable, oil drums, and other ruins of this little polar station.

Mike Fay and I entered one derelict house and found a foot of ice in the corridor, an old floor lamp, a pair of rusty ice skates, and a poster-calendar from 1990 still hanging on one wall, with a legend saying, "All Union Society of Book Lovers." This bibliophile spirit was affirmed in the rec center, a larger building, where someone took the trouble before evacuation to put a padlock on the door to the library; through the grated wall, we could see hundreds of books still standing neatly on their shelves. But the pool table in the outer room was a mess.

"*C'est la fin de l'empire,*" Romanenko said, not complicating his French with past tense. He meant the end of the Soviet Empire. He was old enough to remember.

More than one empire had fallen since an Austro-Hungarian expedition came to these islands in 1873. More than one flag was raised here that no longer flew. More than one geophysical expectation, such as the existence of an Arctic continent, had been debunked. The North Pole is real, as a determinable if invisible point, but the early explorers such as Nansen, who came or went via this archipelago with their dog teams and their ice-riding ships, failed to reach that elusive spot. Franz Josef Land has been a memorable waypoint on the glorious polar route toward frustration and disillusion. Its lonely flat-topped islands, with their parapets of basalt, stand as emblems of frigid adamance; they testify that, though men can be stubborn and resourceful and brave, nature is surpassingly complex and strong. That's the meaning of true north, far north, absolute north.

The remains of old Krenkel Station tempered that testimony to nature's preeminent power in their ambivalent way: with hundreds of tons of industrial garbage and delicate vestiges of the humanity of those who hunkered here.

Because the station was part of Franz Josef Land and because Franz Josef Land was within the administrative ambit of Russian Arctic National Park (though not yet enjoying full park protection itself), park authorities had initiated cleanup operations at Krenkel. They envisioned subsuming the site within a planned *muzey pod otkrytym nebom,* or great open-air museum. But they would face some delicate decisions about where remediation should stop and preservation begin. When such a place lands on the junk heap of history, how do you tell what's history and what's junk?

Even more delicate, and far more consequential, would be decisions made in Moscow about renewed Russian military attention to the Arctic. In early November 2013, just two months after we finished this voyage, Defense Minister Sergei Shoigu announced plans to deploy a squadron of warships with ice-breaking capability to protect new trans-Arctic sea routes as well as potential oil and gas deposits. Protect against what? Interference and preemption by other nations, evidently. As of 2011, according to the Russian news agency Novosti, 95 percent of Russia's natural gas reserves and 60 percent of its oil reserves lay in the Arctic region, although most of the fields were beneath the Barents and Kara Seas, closer to the mainland. The discovery of those fields and the warming climate had encouraged Russia to look farther north. (More recently, in consequence of Russia's war against Ukraine, the prospect of oil and gas development in the Russian Arctic has become fraught; see the update at the end of this piece.) The defense minister's 2013 announcement even mentioned reopening the Nagurskoye air base, on one of the westernmost islands of Franz Josef Land. Would this proprietary surge, if it happened, be compatible with another form of protection, the protection of Arctic ecosystems? Enric Sala, a calm optimist, thought it would. After all, Vladimir Putin himself pretended to harbor conservationist sympathies—but who could tell with Putin? Sala hoped that Franz Josef Land would soon receive full protected status as a national park, and he reckoned that a strengthened military presence "can actually help with enforcement."

THE QUESTION OF ice underlying all these issues will not be answered by any one expedition. Measurements can be taken, photographs can be shot, comparisons can be made between ice coverage now and what early explorers saw, but the matter of causality is vast and intricate. The scientists on this team did what good field scientists always do: They gathered quantitative observations of particulars. Making dive after dive in the freezing water, Alan Friedlander, a marine ecologist from the University of Hawai'i, identified 16 species of shallow-zone Arctic fish and began pondering why diversity here seemed to be low. Kike Ballesteros, a Harvard botanist, likewise spending his days in a dry suit with numbed fingers and reddened cheeks, made a thorough inventory and biomass assessment of the marine algae, something never before done. Maria Gavrilo and her team censused ivory gulls, kittiwakes, guillemots, little auks, common eiders, and glaucous gulls, measuring, weighing, banding, and placing geolocators on some. Forest Rohwer and his graduate student Steven Quistad, from San Diego State University, captured billions of viruses from a variety of hospitable media, such as beach slime and guano, and would derive insights from sequencing their genomes back in Rohwer's lab. Mike Fay identified and collected more than 30 species of flowering plants. Daria Martynova, from the Zoological Institute of the Russian Academy of Sciences, sampled the water column for copepods, gauging the penetration of that North Atlantic species *Calanus finmarchicus* into the Arctic realm of *Calanus glacialis*. Such efforts, and all the other observations gathered during the expedition, would help answer smaller questions within the big one.

Was the planktonic community changing? Were the kittiwakes and guillemots reproducing as successfully as in the past? Had the sea-bottom fauna or the terrestrial flora been affected by trends of temperature change? Had the polar bears become more concentrated on islands, marooned there now that the sea ice had vanished from Franz Josef Land during summer? Had the planktonic changes, if any, adversely affected the population of little auks? This is ecology—everything interconnected. The whole body of data and analyses would be pulled together within coming months into a compendium report under Sala's editorial eye.

Through the end of our journey and beyond, I carried vividly in memory a moment that occurred near the beginning, while I was ashore on Hooker Island with the Frenchmen, David Grémillet and Jérôme Fort. We had spent a long afternoon with their noose carpets deployed, getting only modest results. They had caught and processed three little auks. It wasn't enough data, and at that rate they would need to change their tactics or choose a different site. Then, as Fort and I gathered our gear to depart, Grémillet spotted an adult auk hiding between the boulders, where auks place their nests. He grabbed it. Doing that, he spotted something else: a chick. He grabbed the chick too and turned to us, an auk in each hand. Measuring and banding a bird takes two hands; extracting a blood sample takes four. These two scientists, after a slow day, were suddenly busy. So Grémillet handed me the chick. I accepted it in my cupped hands, with a high sense of privilege, and tried to shield it from the wind.

Little auks have long lives, up to 20 years, and reproduce slowly, one chick a year. Each chick is precious. The period from hatching to fledging, the most vulnerable time in an auk's life, is about 25 days. This chick had just hatched. It was a ball of fluffy black down, the size of a plum, with a beak. Trusting and helpless. After a short time, I passed it gingerly back to Grémillet, and he returned it to its nest.

Recalling the moment months later, back in the U.S., I wondered where that bird was. I wondered whether it survived its 25 days in the rocks, fledged, and flew away from Franz Josef Land to a wintering ground somewhere, an exemplary little auk, intrepid and resilient, representative of the fact that living creatures will find a way, even in the far, far north.

....

This expedition was a success by at least one measure: On August 25, 2016, Russian Arctic National Park was officially expanded to include Franz Josef Land, adding more than 6,000 square miles of land area and 22,000 square miles of sea area to the park, which more than sextupled the total expanse of protected land and sea.

As originally established in 2009, Russian Arctic National Park encompassed only a northern portion of Novaya Zemlya, a long, thin island defining the eastern perimeter of the Barents Sea. Franz Josef Land helps mark the north perimeter of that sea. Just a few years after our journey through FJL, Vladimir Putin visited the place too—probably a flying trip to the Nagurskoye air base, the only site where a presidential plane could land—and on that basis described the archipelago as "a giant rubbish tip." He must have seen the wrecked and abandoned trucks, radar equipment, derelict planes, and other military hardware, as well as the quarter million barrels of oil products left behind when the wheels fell off the Soviet Union in 1991. That debris was the Soviet Air Force's equivalent to what we saw at Krenkel Station, only more so. Putin may not have had time to admire the polar bears and walruses, let alone the ivory gulls, kittiwakes, guillemots, and little auks, except perhaps from a helicopter. But he didn't despair of the place, evidently, and two years later his government undertook an expensive cleanup effort (running to hundreds of millions of rubles) intended to make Franz Josef Land fit for tourism. Conservation can seem appealing, even to a cynical autocrat like Putin, when there's money in it.

As for Russian development of oil and gas deposits elsewhere in the country's Arctic ambit: Those ambitions seem jeopardized by international sanctions imposed in response to Putin's war of aggression against Ukraine. Even before this war, some Western energy companies backed off their involvement in Russian Arctic work after the grabbing of Crimea in 2014; then in 2022, when Putin's government became anathema, the trend worsened. Both access to special technology and financial participation by Western firms were at issue. "It will be extremely hard for the Russians to develop oil and gas resources without Western technology," one industry expert, David Goldwyn of Goldwyn Global Strategies, told National Public Radio in April 2022. "They have some conventional drilling technology themselves, but not the special coatings and pipe and artificial intelligence, which they need to develop those oil and gas fields." It could mean delay for some projects, cancellation or indefinite mothballing for others.

These are countercurrent trends. The addition of Franz Josef Land to Russian Arctic National Park reflects an expectation that some hardy travelers, especially those with a deep fondness for seabirds and white bears, might eventually—if and when it again seems imaginable for any Western tourist to visit any part of Russia—pay serious money to explore its northernmost islands. I hope that time comes, and I hope the tourists do too (but see "Let It Be," pages 165–172), so long as such visitors can confine their sleeping and eating and other direct impacts to offshore boats (as is required of tourists in the Galápagos Islands) and step lightly when they go ashore.

Enric Sala told me recently that the old air force base, Nagurskoye, has even been revivified and expanded, with new buildings and a bigger runway. The cheerful way of seeing this, he added, is that military presence should be a disincentive to poachers. The less cheerful way, it occurs to me, is that Putin's government is grinning with shiny new steel teeth.

As for Pristine Seas, Sala's project: Its expeditions have continued and yielded further results for marine conservation, largely in tropical waters. In 2014, he and a team made survey dives near remote U.S.-territory islands in the western Pacific, between Wake and Kiribati, after which Pacific Remote Islands Marine National Monument (under management of the U.S. Department of the Interior) was expanded to its current size, almost a half million square miles of open ocean, coral reefs, and island habitats. They also surveyed waters near Palau and confirmed for the Republic of Palau that its system of traditional temporary fishing closures, known as buls, is indeed functioning effectively to protect populations of reef fishes. In 2017, they dove off Gabon, helping that country gather data for its Gabon Bleu program, Africa's largest system of marine protected areas, which encompasses, among other things, important humpback whale nurseries. And in 2018, Sala and company discovered a new hydrothermal vent field off the Azores.

Altogether, since its founding in 2008 with support from the National Geographic Society, Pristine Seas has assisted in creating 26 ocean reserves, which protect extraordinary marine ecosystems, from the kelp forests off Tierra del Fuego to the shark strongholds of the Galápagos to the offshore waters of Franz Josef Land. Practically speaking, such protections enhance the sequestration of carbon in undisturbed marine sediments—and carbon sequestration is, of course, crucial to retarding climate change. Further, marine protected areas serve as abundant sources of larval fish, adult fish, and invertebrates that go forth and repopulate unprotected areas where fishing is allowed—a boon to local economies.

Sala and his colleagues are ambitious to do more. With 2.5 percent of the total oceanic area now protected, they are touting the potential benefits, to humans as well as to marine biodiversity, of achieving 35 percent protection. To reach that goal will take a lot of diving, data gathering, and patient dryland diplomacy. It will also require a lot of support from citizens and leaders awakened to the reality that Earth is known as the blue planet—not the brown planet or the gray planet or even the green planet—for a reason.

RESERVOIR
OF TERROR

∗ · ∗ · ∗

Ebola Virus Doesn't Disappear—
It Just Hides

....

The West African epidemic of Ebola virus disease, in 2013–16, was the last instance of a sudden and catastrophic viral disease event preceding the one we call COVID-19. Just as the pandemic-causing coronavirus, now known as SARS-CoV-2, was referred to as "a novel virus" in the early days of January 2020, Ebola virus itself was once novel—back in 1976, at the time of its first recognized outbreak in humans. Because it was newly identified then as an infection of humans, and because everything comes from somewhere, the question of Ebola's origin presented itself, from the start, as both mystifying and urgent.

Ebola came to be included in a group of novel and nefarious pathogens loosely known as "emerging viruses." But from where did this virus emerge? Stated otherwise: In what creature does it lurk, inconspicuously, without causing notable symptoms, when it's not killing people? That creature (safe to assume it's an animal because new human viruses typically come from other animals) is referred to (as I explain in the story) as its reservoir host. The reservoir host is a long-term refuge for the virus, which has adapted over centuries or millennia to be relatively innocuous to that host, which provides a stable hideaway. A virus sometimes spills from its reservoir host into humans during close contact between people and wild animals. Often the contact occurs because of habitat destruction, killing or capture of wild animals for food, or other kinds of ecological disruption.

Identifying that moment of fateful contact, and preventing further such events in the future, requires learning the identity of the reservoir host. Ebola's host was still unknown in 2015, when I set off on this assignment, and remains unknown today. The matter of Ebola's origin is not just an abiding question but a resonant one in the context of COVID-19, because the origin of SARS-CoV-2 is still being investigated, and contentiously argued, as of this writing.

I've been fascinated by the ecological dimensions of emerging viruses generally, and of Ebola virus in particular, for about 25 years. During the horrible West Africa event, National Geographic *asked me to do an Ebola story. Other journalists were covering the medical crisis and the public health issues in Liberia, Guinea, and Sierra Leone; I might be more useful, I reckoned, by focusing on the virus itself. At almost the same time, I heard from my friend Fabian Leendertz, a veteran disease ecologist in Berlin, that he and a team of students were headed for Ivory Coast, in West Africa, to explore a hypothesis about Ebola's reservoir host. I tagged along, in partnership with Pete Muller, a seasoned conflict photographer.*

No one foresaw, back in December 2013, that the little boy who fell ill in a village called Méliandou, in Guinea, West Africa, would be the starting point of a gruesome epidemic, one that would devastate three countries and provoke concern, fear, and argument around the planet.

No one imagined that this child's death, after just a few days' suffering, would be only the first of many thousands. His name was Emile Ouamouno. His symptoms were stark—intense fever, black stool, vomiting— but those could have been signs of other diseases, including malaria. Sad to say, children die of unidentified fevers and diarrheal ailments all too frequently in African villages. But soon the boy's sister was dead too, and then his mother, his grandmother, a village midwife, and a nurse. The contagion spread through Méliandou to other villages of southern Guinea.

This was almost three months before the word "Ebola" began to flicker luridly in email traffic between Guinea and the wider world.

The public health authorities based in Guinea's capital, Conakry, and the viral disease trackers from abroad weren't in Méliandou when Emile Ouamouno died. Had they been, and had they understood that he was the first case in an outbreak of Ebola virus disease, they might have directed some timely attention to an important unknown: How did this boy get sick? What did he do, what did he touch, what did he eat? If Ebola virus was in his body, where did it come from?

Among the most puzzling aspects of Ebola virus, since its first recognized emergence in 1976, is that it disappears for years at a time. Since that first outbreak in what was then Zaire (now the Democratic Republic of the Congo) and a simultaneous episode with a closely related virus in what was then southern Sudan (now the country South Sudan), the sequence of Ebola events, large and small, has been sporadic. During one stretch of 17 years (1977–1994), not a single confirmed human death from infection with Ebola virus occurred. This is not a subtle bug that simmers delicately among people, causing nothing more than mild headaches and sniffles. It kills. If it had been circulating in human populations for those 17 years, we would have known.

A virus can't survive for long, or replicate at all, except within a living cellular creature. That means it needs a host—at least one kind of animal, plant, fungus, or microbe whose body serves as its primary environment and whose cell machinery it can co-opt for reproducing. Some harmful viruses dwell in nonhuman animals and only occasionally spill into people. They cause diseases that scientists label zoonoses. Ebola virus disease is a zoonosis, an especially nasty and perplexing one, killing many of its human victims in a matter of days, pushing others to the brink of death, and then vanishing. Where does the virus hide, quiet and inconspicuous, between outbreaks?

Not in chimpanzees or gorillas; field studies have shown that Ebola virus often kills them too. Dramatic die-offs of chimps and gorillas have occurred around the same time and in the same area as Ebola virus outbreaks in humans, and some carcasses have tested positive for signs of the virus. Scavenging ape carcasses for food, in fact, has been one of the routes by which humans have infected themselves with Ebola. So the African apes are highly unlikely to harbor Ebola chronically. It hits them and explodes. It must lurk somewhere else.

The creature in which a zoonotic virus exists over the long term, usually without causing symptoms, is known as a reservoir host. Monkeys serve as reservoir hosts for the yellow fever virus. Asian fruit bats of the genus *Pteropus* are reservoirs of Nipah virus, which killed more than a hundred people during a 1998–99 outbreak in Malaysia. Fruit bats also host Hendra virus in Australia, where it drops from bats into horses, with devastating effect, and then into horse handlers and veterinarians, often killing them. The passage event, when a virus goes from its reservoir host to another kind of creature, is termed "spillover."

As for the reservoir host of Ebola—if you have heard that fruit bats again are the answer, you've heard supposition misrepresented as fact. Despite arduous efforts by some intrepid scientists, Ebola virus has never been tracked to its source in the wild. The gold standard for identifying a reservoir host is to take a sample of blood or other material from the suspected creature and culture the live virus—grow it—in a lab. That hasn't happened with this most notorious of the ebolavirus group, the one called *Zaire ebolavirus,* or simply Ebola.

"Where is it when it's not infecting humans?" Karl M. Johnson posed that question, to himself more than to me, during one of our conversations. Johnson is an eminent virologist, a pioneer in Ebola virus research, the former head of the Viral Special Pathogens Branch at the Centers for Disease Control and Prevention (CDC). He led the international response team against that initial 1976 outbreak in Zaire, a harrowing venture into the unknown. He also led a team that isolated the virus in a CDC lab, demonstrated that it was new to science, and named it after a modest Zairean waterway, the Ebola River. Johnson wondered back then about its hiding place in the wild. But the urgency of human needs during any Ebola outbreak makes investigations in viral ecology difficult and unpopular. If you're an African villager, you don't want to see foreigners in moon suits methodically dissecting small mammals when your loved ones are being hauled away in body bags. Thirty-nine years later, when I spoke with him that day, Johnson noted that the identity of the reservoir host was "still largely a monster question mark out there." And it still is.

In April 2014, soon after word spread that the cluster of deaths in southern Guinea involved Ebola virus, Fabian Leendertz arrived there with a team of researchers. Leendertz is a German disease ecologist and veterinarian, then based at the Robert Koch Institute in Berlin, who studies lethal zoonoses in wildlife, with special attention to West Africa. (Later, in 2021, he became founding director of the Helmholtz Institute for One Health in Greifswald, Germany.) Leendertz reached southern Guinea by driving overland from Ivory Coast, where he had worked for 15 years in Taï National Park on disease outbreaks among chimpanzees and other animals. He brought with him three big vehicles, full of equipment and people, and two questions. Had there been a recent die-off among chimps or other wildlife, possibly putting meat-hungry humans at risk from infected carcasses? Alternatively, had there been direct transmission from the Ebola reservoir host, whatever it was, into the first human victim? Leendertz knew nothing at that point about Emile Ouamouno. His team spoke with officials and local people and walked survey transects through two forest reserves, finding neither testimony nor physical evidence of any remarkable deaths among chimpanzees or other large mammals. Then they shifted their attention to the village of Méliandou, talked with people there, and heard a very interesting story about a hollow tree full of bats.

These were small bats, the dodgy kind that echolocate and feed on insects, not the big creatures that fly out majestically at dusk to eat fruit, like a Halloween vision of nocturnal crows. The locals called these little bats *lolibelos*. They were dainty as mice and smelly, with wriggly tails that extended beyond their hind membranes. Showing pictures and taking descriptions, Leendertz's team ascertained that the villagers were probably talking about the Angolan free-tailed bat *(Mops condylurus)*. Great numbers of these bats had roosted within a big, hollow tree that stood beside a trail near the village. Then, just weeks before, the tree had been burned, possibly during an attempt to gather honey. From the burning tree came what the people remembered as "a rain of bats." The dead bats were gathered up, filling a half dozen hundred-pound rice sacks, and might have been eaten except for a sudden announcement from the government that, because of Ebola, consuming bush meat was now prohibited. The Méliandou villagers threw the dead bats away.

And there was something else about that hollow tree, the villagers told Leendertz's team. Children, possibly including Emile Ouamouno, used to play in it, sometimes catching the bats. They would even roast them on sticks and eat them.

Leendertz consulted a colleague with expertise in recovering DNA from environmental samples, who told him it might be feasible to find enough beneath the tree to identify the bat species that had roosted there. "So I started running around with my tubes and spoon collecting soil," Leendertz told me. Back in Berlin, genetic sequencing confirmed the presence of Angolan free-tailed bats. On that evidence, this creature—an insectivorous bat, not a fruit bat—joined the list of candidates for the role of Ebola's reservoir host.

THE FIRST CLUES in the long mystery—clues that seemed to point toward bats—arose from disease outbreaks caused by Marburg virus, Ebola's slightly less notorious relative within the family known as filoviruses. The story of Ebola is closely connected with that of Marburg, according to a seasoned South African virologist named Robert Swanepoel, who has long studied them both.

"The two are interlinked," he said, as we sat before a computer screen in his Pretoria home, looking at photographs from his archive. Swanepoel, who hides a genial heart within a bearish exterior, had retired from the National Institute for Communicable Diseases (NICD), in Johannesburg, where he ran the Special Pathogens Unit for 24 years, but he was still busy with varied work and bristling with ideas and memories.

Back in 1967, nine years before Ebola itself was recognized, a shipment of Ugandan monkeys intended for medical research arrived in Frankfurt and Marburg, West Germany, and in Belgrade, Yugoslavia, bringing with them an unknown but dangerous virus. Laboratory workers became infected in each place, and then, secondarily, some family members and health workers. Among 32 confirmed cases, seven people died. The new virus, a spooky thing, filamentous as seen by electron microscopy, like a strand of toxic vermicelli, was given the name Marburg virus. Eight years later an Australian student died of Marburg virus disease in a Johannesburg hospital after a hitchhiking trip across Rhodesia (now Zimbabwe). He and his girlfriend—she got sick too but recovered—had done a few things that might have exposed them to infection: slept on the ground in a pasture, bought some raw eland meat, fed some caged monkeys. And they had visited the Chinhoyi Caves, a complex of caverns and sinkholes in northern Rhodesia that, like many caves in Africa, have been known to harbor bats. Along the way the hitchhiker also sustained some sort of insect or spider bite, which raised a painful red welt on his back. Investigation of his case in the immediate aftermath focused much on the bite, little on the caves.

Two other early cases of Marburg virus disease did cast some suspicion on caves and the bats that roost within them. In 1980, a French engineer who worked at a sugar factory near the base of Mount Elgon, in western Kenya, ventured into Kitum Cave, a deep passage into the volcanic rock of the mountain, sometimes entered by elephants looking for salt. The engineer's cave visit was evidently a bad idea; he died of Marburg in a Nairobi hospital. In 1987 a Danish schoolboy climbed the mountain and explored the same cave during a family vacation, and he died of an infection with a virus (now known as Ravn virus) closely related to Marburg. These events engaged the notice of Swanepoel, down in Johannesburg. In 1995 came another outbreak—Ebola this time, not Marburg—centered on the city of Kikwit in what is now the Democratic Republic of the Congo (DRC). The chain of human-to-human infections, which totaled 315 cases and 254 deaths, began with a man who farmed manioc and made charcoal in a forest area at the city's edge. Swanepoel flew to Kikwit, joining an international team of responders. He came down with malaria, went home, recovered, and in early 1996, with the support of the World Health Organization, returned. His primary task was to look for the reservoir host, searching the same ecosystem where the outbreak had begun at the same time of year. "Already by that stage," he told me, "bats were on my mind."

Swanepoel and his crew at Kikwit took blood and tissue not only from bats but also from a wide selection of other animals, including many insects. Screening those samples back at his lab in Johannesburg, he found no evidence of Ebola. So he tried an experimental approach, one that seemed almost maniacally thorough. Working in NICD's high-containment suite—biosafety level 4 (BSL-4), the highest—he personally injected live Ebola virus from the Kikwit outbreak into 24 kinds of plants and 19 kinds of animals, ranging from spiders and millipedes to lizards, birds, mice, and bats, and then monitored their condition over time. Though Ebola failed to take hold in most of the organisms, a low level of the virus—which had survived but probably hadn't replicated—was detected in a single spider, and bats sustained Ebola virus infection for at least 12 days. One of those bats was a fruit bat. Another was an Angolan free-tailed bat, the same little insectivore that would later catch Fabian Leendertz's attention in Méliandou. It was proof of principle, though not of fact: These creatures could be reservoir hosts.

THE EVENTS IN Kikwit highlighted an important difference between Marburg and Ebola viruses that has persisted: Whereas outbreaks of Marburg virus disease usually begin around caves and mines, Ebola virus disease outbreaks usually begin with hunting and carcass scavenging, which are forest activities. This suggests the two viruses may emerge from two different kinds of reservoir hosts—or if bats are the hosts, two different kinds of bats, cave roosters and tree roosters.

The pattern was reaffirmed during a cluster of Marburg outbreaks from 1998 to 2000, centered on a derelict gold-mining town called Durba, in the DRC. Bob Swanepoel led another expedition and found multiple chains of infection, most or all of which started with miners who worked underground. Miners who worked at open pits in the daylight were far more likely to stay healthy. This led Swanepoel to suspect cave-roosting Egyptian fruit bats as the virus source, though he didn't publish his suspicion at the time.

Then, beginning in late 2001 and extending into 2003, another series of small, independent outbreaks—of Ebola again, not Marburg—afflicted villagers in the densely forested borderlands of Gabon and the Republic of the Congo (which are west and north of the DRC, on the other side of the Congo River). Roughly 300 people became infected; almost 80 percent died. Meanwhile, gorillas, chimpanzees, and duikers started turning up dead in the same region. Each human outbreak seemed to start with an unfortunate person, usually a hunter, who had handled an animal carcass.

"People were dying, and different animals were dying," said Janusz Paweska, nowadays Swanepoel's successor as head of Special Pathogens at NICD, when I visited him in Johannesburg. "So we thought, this is a good time to hunt for the Ebola reservoir."

Swanepoel enlisted Paweska and others, then arranged a partnered expedition with Eric Leroy, a French virologist based in Gabon who had responded to earlier Ebola outbreaks there. He met with Leroy in Gabon's capital, Libreville, before heading into the field. "I gave him a long story about how historically bats have been involved in Ebola and Marburg," Swanepoel told me. His team, he informed Leroy, had found fragments of Marburg genome, for instance, in the underground bats at Durba. Swanepoel had brought rodent traps, mist nets, and other collecting gear to Gabon. "Although I was fixated on bats, I said we had to cover everything," he recalled. That would include a variety of mammals, birds, mosquitoes, biting midges, and other insects. Swanepoel's group took home a third of the specimens and sent a third to the CDC in Atlanta, leaving a third to be tested by Leroy. The processing moved slowly in Swanepoel's lab and at the CDC, amid many other projects, and yielded no positives. "We drew a blank."

But Leroy's group went back. Eventually his team made three field trips to the border area, capturing and sampling more than a thousand animals, including 679 bats, on which Leroy too was now fixated. In 16 of those bats, belonging to three different fruit-eating species, they found antibodies—proteins afloat in the blood, marshaled by the immune system to battle invaders—that had reacted against Ebola virus. In 13 other fruit bats they detected very short fragments of Ebola RNA. It's important to note that those two kinds of evidence, antibodies and viral fragments, are analogous to finding the footprints of a yeti in snow. You might or might not have something real. Isolating live virus—that is, growing fresh and infectious Ebola from a sample of tissue or blood or feces—is the higher standard of evidence, almost like finding a real yeti's foot attached to a real yeti in a leghold trap. Leroy's group didn't succeed in growing live virus from any samples. Still, in 2005 the journal *Nature* published a paper on these results, written by Leroy but with Swanepoel and Paweska credited as co-authors, titled "Fruit Bats as Reservoirs of Ebola Virus." That paper, though cautious and provisional, was the primary source for most of those careless, overly certain assertions seen in the popular media during 2014 and 2015 to the effect that Ebola virus resides in fruit bats.

Possibly it does. Or not. The paper itself says maybe.

"You tried to isolate live virus?" I asked Leroy during a stopover in Gabon. He was a courteous, dapper Frenchman, serving then as director of the Centre International de Recherches Médicales de Franceville, who worked in a white shirt and dark tie, at least when he wasn't wearing a full protective suit in his BSL-4 lab or Tyvek coveralls in the forest. "Yes. Many, many, many times trying to isolate the virus," he said. "But I never could. Because it was—the viral load was very, very low." Viral load is the quantity of virus in the solid tissues or blood of a creature, and it tends to be much lower in a reservoir host than in an animal or person suffering an acute infection.

That's just one of three reasons why finding a reservoir host is difficult, Leroy explained. The second is that, in addition to low viral load within each animal, the virus may exist at low prevalence within a population. Prevalence is the percentage of positive individuals at a given time, and if that happens to be as little as one animal in a hundred, then "the probability to detect and to catch this infected animal is very low." If a single kind of animal amid the great diversity of tropical forests represents a needle in a haystack, then one infected individual within one population of animals amid such diversity represents one needle in 10,000 haystacks.

And the third constraint on the search for a reservoir host? "It's extremely expensive," Leroy said.

THE COST OF field operations in remote forest locations, as well as the competing demands upon institutional resources, has hindered even veteran researchers such as Swanepoel and Leroy from mounting long-term, continuous studies of the Ebola reservoir question. Instead, there were short expeditions, organized quickly during an outbreak or just as a crisis was ending. Another hindrance, quite legitimate, was human sensitivities. Going to the site of an outbreak to do research on the ecology of the virus is not just logistically nightmarish but, as I've mentioned, offensive to local people. So those expeditions get delayed. The problem with delay is that the prevalence of Ebola virus within its host population, the viral load within individual hosts, and the abundance of virus being shed into the environment may all fluctuate seasonally. Miss the right season, and you might miss the virus.

Fabian Leendertz tried to address these difficulties by organizing a second field expedition, this one at roughly the same season as the fateful spillover that killed Emile Ouamouno, but a year later and in neighboring Ivory Coast. Angolan free-tailed bats were abundant there too, roosting beneath the roofs of village houses. Their very abundance in such proximity to people suggested a further perplexing question, if the little-bat hypothesis was correct: With the virus so near, why didn't spillovers occur far more often? Leendertz wanted to trap those bats, as many as possible, and sample them for evidence of Ebola. Photographer Pete Muller and I went with him.

Leendertz and his team, including a graduate student named Ariane Düx, focused on two villages outside the city of Bouaké, a trade hub near the country's center. After shopping for trap materials in Bouaké's market, scouting the villages for bat-filled houses, and paying respects to village elders, the team assembled their apparatus late one afternoon, in time for the fly-out at dusk. The traps were cone-shaped structures, jerry-built of long boards and translucent plastic sheeting, designed to capture bats as they emerged from a roof hole and funnel them down into a plastic tub. Amazingly, the system worked. At 6:25 p.m. on the first evening, one trap came alive like a popcorn popper, as dozens of small gray bodies slid down the sheeting and thumped into the tub.

For the next phase, Leendertz and Düx suited up in medical gloves, respirator masks, gowns, and visors. With a naked lightbulb hanging above their makeshift lab table, they began processing bats: weighing and measuring each animal, noting sex and approximate age, injecting an electronic chip the size of a caraway seed for later identification, and most important, drawing blood from a vein in the animal's tiny arm. One well-aimed poke with a delicate needle, and a blood drop would appear, to be gathered with a fine pipette. Düx and Leendertz worked together at close range, trustingly sharing tasks, and it occurred to me that if she poked twice at the vein and missed the second time, jabbing Leendertz's finger instead, he could have an Ebola-related needle-stick injury. But she didn't miss.

The blood went into small vials, for freezing immediately in a liquid-nitrogen tank and eventual screening back in Berlin. A small fraction of all the captured bats would be killed and dissected, so that snippets of their internal organs, especially liver and spleen, where viruses often concentrate, could be added to the trove of frozen samples. The other bats would be released. If a blood sample from one dissected individual later tested positive for antibodies or viral fragments, its organs would then be used in an attempt (more dangerous and more expensive, done only in a BSL-4 laboratory) to isolate live Ebola virus.

After a few bats, Leendertz stepped back from the processing work and allowed an Ivorian graduate student, Kouadio Léonce—tall, mild-mannered, and thin as a candle—to take his place. This was a training mission as well as a scientific investigation, after all, and Leendertz wanted to give his protégés a richness of experience. Kouadio had good skills already, and as he got into rhythm, sharing these exacting tasks in the warm African night, I noticed the T-shirt beneath his medical gown, which carried some sort of resort logo and said, IT'S THE PERFECT HOLIDAY. For him, maybe, but not for everybody.

Back in the United States, I spoke with more experts during a stop at the CDC in Atlanta and by telephone. When I asked why it's important to identify the reservoir host of Ebola virus, they all agreed: because that information is essential to preventing future outbreaks. On other points they diverged. The most unexpected comment came from Jens Kuhn, a brainy young virologist at the National Institutes of Health and, by way of his tome *Filoviruses,* arguably the preeminent historian of Ebola. I've known Kuhn as a candid source but also a lively and generous friend since we met years ago at a conference in Libreville hosted by Eric Leroy. Why do you think that, after all these years, I asked him, the reservoir of Ebola is still unidentified?

"It's a strange host."

"A strange host," I repeated, not sure I'd heard right.

"That's what I think."

His logic was complex, but he sketched it concisely. First, outbreaks of Ebola virus disease had been relatively infrequent—only about two dozen, at that point, in nearly 40 years. Rare occurrences. Almost every one of those was traceable to a single human case, infected from the wild, followed by human-to-human transmission. This suggests, Kuhn said, that the sequence of events yielding spillover must be "extraordinary and weird." Highly unusual circumstances, an unlikely convergence of factors. Second, there's "the remarkable genome stability of the virus over the years." It didn't change much, didn't evolve much, at least until the human case count in West Africa started going so high, providing many more opportunities for the virus to mutate. That stability might reflect "a bottleneck somewhere," Kuhn said—a constraining situation that keeps the virus scarce and its genetic diversity low. One possible form of bottleneck would be a two-host system: a mammal host such as a bat that becomes infected only intermittently, when it gets bitten by a certain insect or tick or other arthropod, perhaps relatively rare or narrowly distributed, which is the ultimate host of the virus. As we both knew, this harked back to 1975 and that hitchhiker in Rhodesia who suffered an odd little bite and then died of Marburg. It evoked the spider in Bob Swanepoel's lab that carried Ebola for two weeks.

What would you do if you had a big research grant for nothing but finding Ebola's reservoir? I asked Kuhn. He laughed.

"I'm going to make myself unpopular," he said, "but I would still look into insects and other arthropods."

He doesn't have that big grant, nor does anyone else. The mystery abides. The stakes are high. The samples from Leendertz's mission to Ivory Coast yielded no positives. The search continued.

· ·

And it continues still. Ebola remains a mysterious disease as well as a horrific one. The epidemic in West Africa has been unmatched, in the annals of Ebola outbreaks, for sheer scope of misery and death— more than 28,000 cases, more than 11,000 deaths, in those three unfortunate countries, plus a very few victims who carried the virus to Nigeria, Mali, the United States, and elsewhere. But in the years since that ended, more than a half dozen further outbreaks of Ebola virus disease have occurred, most of those in the Democratic Republic of the Congo. Even as I write this, there are reports of new Ebola fatalities in DRC's northwestern province of Équateur, in and around the city of Mbandaka. The case numbers are low, so far, which may reflect a strain of the virus that is less transmissible, or else just good public health measures in action. There is now a vaccine against Ebola virus disease, which can be delivered quickly to residents of areas at risk. The fatality rate among confirmed cases in this new outbreak, on the other hand, seems unusually high (and with Ebola, it's always fearfully high), which may suggest an exceptionally virulent strain. These are informed speculations, but just speculations, and the time is early; the hard data are scarce. Mbandaka is a remote and underserved city.

The longtime and recurrent associations of Ebola outbreak events with the Democratic Republic of the Congo, one of the world's greatest national repositories of biological diversity, and of tropical forest in particular, is not likely coincidental. Almost all evidence points to the likelihood that Ebola is a forest and village disease, specifically of equatorial Africa and mostly recurrent within countries of Central Africa: DRC, the Republic of the Congo, Gabon, Uganda, and what is now South Sudan. The West African epidemic was devastating—by far the biggest Ebola event ever—but geographically anomalous relative to the historical pattern, and it's important to remember that disaster seems to have originated from a single spillover into a single human, probably the little boy in Méliandou village. The event may have begun with close contact between a human and a bat, though not necessarily an Angolan free-tailed bat. Or it may not have involved a bat. Other outbreaks, such as the one in a Gabonese village called Mayibout 2, in 1996, or the one centered in a district called Mbomo, in the Republic of the Congo, during 2003–04, seem to have started with contact between a human and the carcass of a chimpanzee or a gorilla (as I described in "The Green Abyss," pages 41–62). Chimps and gorillas are absolved of suspicion as reservoir hosts of the virus, though, as I've mentioned, because they too sicken and die from it. Their luckless role seems to be as intermediaries.

Until the reservoir host of Ebola is identified and confirmed with that gold-standard form of evidence—culturing the virus, in its full functionality, from some nonhuman animal—this formidable virus will reside in a sort of black box of human unknowing. All we can say for sure is that it's out there, somewhere, in the great Central African forests, and we disturb it at our peril.

SAVING THE OKAVANGO

⬩⫝⬩⫝⬩

A Great Wetland Needs Water

....

Steve Boyes came to maturity as a biologist and conservationist in the first years of the 21st century, just the right time to absorb deeply the inspirational model of Mike Fay and his Megatransect: Think big, bury your naked toes in the landscape you hope to protect, travel through it with your eyes and pores open, carry a flame in your heart, let others consider you a fanatic, but be pragmatic as well, and steel your will for a long struggle.

Boyes made that linkage explicit when he titled his 2015 expedition the "Source to Sand Megatransect." He and his team traveled by canoe from the headwaters of the Cuito River in central Angola to the Okavango Delta in Botswana and, beyond that, the Kalahari Desert. This journey of discovery was only part of a multiyear effort to study and document the waters, the life in those waters, the circumstances of use and human demand upon those waters, that make the Okavango what it is. When they returned to Angola in early 2017, to conduct a parallel exploration along the Cubango River, which meets a confluence with the Cuito to form the Okavango River at the Namibian border, National Geographic *asked me to tag along.*

Having walked with Fay himself during parts of the original Megatransect, I was a little skeptical of Boyes's explicit emulation, but also keen to see these unique stretches of riverscape and landscape over the shoulder of a naturalist so expert and impassioned. I was not disappointed.

....

S een from space, high above Africa, the Okavango Delta resembles a gigantic starburst blossom pressed onto the landscape of northern Botswana, its stem angling southeastward from the Namibian border, its petals of silvery water splayed out for a hundred miles across the Kalahari Basin. It is one of the planet's great wetlands, a vast splash of life-nurturing channels and lagoons and seasonal ponds amid a severely dry region of the continent.

This delta doesn't open to the sea. Contained entirely within the basin, it comes to a halt along a southeastern perimeter and disappears into the deep Kalahari sands. It can be thought of as the world's largest oasis, a wet refuge supporting elephants, hippos, crocodiles, and African wild dogs (also known as painted wolves); lechwe and sitatungas and other wetland antelopes; warthogs and buffalo, lions and zebras, and birdlife of wondrous diversity and abundance—not to mention a tourism industry worth hundreds of millions of dollars annually. But from high in space, you won't see the hippos on their day beds. You won't see the wild dogs hunkered in shade beneath thorn scrub, or the glad expressions on the faces of visitors and local entrepreneurs. Another thing you won't see is the source of all that water.

The water comes almost entirely from Angola, Botswana's complicated neighbor, two countries away. It begins in the moist highlands of Angola's rainy center and flows toward the country's southeast—flows quickly in one major drainage, the Cubango, and more slowly in another, the Cuíto, toward which it percolates slowly from source lakes, through grassy floodplains and peat deposits, then seeps into tributaries and from them to the river. The Cuíto and Cubango Rivers converge at the southern Angolan border, forming a bigger river, the Okavango, which flows across the Caprivi Strip, a narrow band of Namibia, and into Botswana. On average, in a year, 2.5 trillion gallons of water arrive.

Take away that liquid gift, rendered continuously by Angola to Botswana, and the Okavango Delta would cease to exist. It would become something else, and that something would not include hippos, sitatungas, or African fish eagles. If southern Africa were a vast golf course, Okavango with the faucets closed would be one of its sand traps.

Changes now occurring or foreseeable in southeastern Angola—in land use, water diversion, population density, and commerce—make this dark prospect a real possibility. That's why the Cuíto and Cubango Rivers, two remote waterways, have quietly attained high interest in certain circles. That's why an international group of scientists, government officials, resource planners, and hardy young explorers, brought together by a fervent South African conservation biologist named Steve Boyes, with support from the National Geographic Society, has embarked on a grand effort of exploration, data gathering, and conservation advocacy called the Okavango Wilderness Project. These collaborators recognize that the well-being and future of the Okavango Delta is at stake—and that the well-being and future of southeastern Angola, a hard landscape, a poor cousin to glorious Okavango, is at stake too.

"We're on borrowed time," Boyes told me, as we sat at a campsite along the Cubango River after a long day of paddling our mokoros (Okavango-style canoes) downstream. Having grown up in Johannesburg, with a passion for nature, Boyes worked for years at various jobs—a bartender at wineries, a naturalist and guide, a camp manager in the Okavango Delta. Along the way he finished a doctorate. By 2007 he had become acutely aware of the water-source issue and tried to raise the alert among people of Botswana, but he mostly met fatalism.

"They were just not interested," he said, recalling a typical reaction: *Yeah, Angola is such a terrible, bad place, and it's such a shame the river may die.* That goaded him to action. He began looking north, toward the headwaters. "We are going to do this," he vowed. "We're going to try and understand what this system is about." In fact, he hoped not just to understand it but also to help preserve it.

ANGOLA IN 2017 seemed an unlikely site for visionary conservation efforts, yet it offered unusual opportunities. It had been ravaged by war but now enjoyed peace. From the early 1960s until the start of the new millennium, Angola was high on the list of nations you would not want to visit, unless you were a mercenary soldier or a diamond buyer. Once a Portuguese colony, it got its independence in 1975, after a bloody war of liberation, then was wracked by civil war for 27 years, a proxy battleground for the superpowers, pustulated with land mines, a scene of great suffering and strife.

But things had changed drastically since 2002, when the rebel party, UNITA, suffered a crushing defeat, after which oil in great quantities began flowing for export and business boomed. "The most important thing we have to tell the world is that Angola is now a stable country," the minister of the environment, Maria de Fátima Monteiro Jardim, told me in June 2017, at a gathering in Luanda, the capital. "We are committed to preserving nature," she said. What that commitment would mean to reality on the ground was a crucial unknown.

The Boyes team had the blessing of Angolan officialdom, along with international support, to pursue an extraordinarily ambitious study of the Cuíto and Cubango Rivers, exploring every mile of them and some of their tributaries, surveying their wildlife, sampling water quality, noting human presence and impacts along the banks, creating a vast and publicly accessible body of data, and trying to comprehend just how the clean waters of southeastern Angola vivify the Okavango Delta in Botswana.

These survey expeditions, eight in total at that point, had been arduous as well as thorough. The first began on May 21, 2015, when Boyes and his team, traveling with an escort of Land Rovers from HALO Trust, the international de-mining organization, and a big Russian cargo truck, arrived at the source lake of the Cuíto River. They had brought several tons of gear and seven mokoros in which to ferry themselves and their stuff downstream. After paddling the length of the lake on their first day, they discovered that the Cuíto at its outlet is a tiny stream, waist-deep but only a yard or so wide, and impossible for navigation by 20-foot-long mokoros. So they dragged the loaded boats downstream, slogging through high grasses alongside the little band of water, pulling like human oxen, and taking data as they went. These mokoros were fiberglass-and-wood models, not dugouts of ebony or some other tree like the Okavango originals, but were still very weighty when fully loaded. Boyes and his team dragged them each day for more than a week before the Cuíto became navigable. Then they climbed aboard, with paddles and poles, but faced a new sort of challenge: crocodiles and hippos.

The Cuíto along its upper reaches is essentially a wilderness river, with clear water, banks lined by reeds, no villages, few signs of humans. On the morning of July 11, 2015, along a broad curve, something plunged through the reeds and into the water just ahead. Boyes, steersman in the lead boat, hollered "Croc!"—a relatively routine alert. He ruddered toward the mid-river channel, giving the animal space along the bank.

Suddenly a great bulge of water rose beside Boyes's boat, as a distraught hippo surfaced—probably a young male, Boyes reckoned by hindsight. Turns out the right evasive line for a crocodile is the wrong one for a hippo. Hippos own the deep water. And like crocs, they kill many people, sometimes hundreds in a year. "It was a big mistake," Boyes told me later. "Completely our fault. We went right over the animal, defending itself."

The hippo drove its lower canine teeth (maybe a foot and a half long, and sharp) through the bottom of the boat. The upper jaw didn't quite catch the gunwale, so instead of biting the mokoro in half, the hippo just capsized the thing, sending Boyes and his bow paddler, a stolid and reliable Brit named Giles Trevethick, into the water. They clambered onto the hull, and a crew member quickly fired a bear-banger flare, meant to disrupt the attack. Boyes's younger brother Chris, his expedition chief, shouted "Swim!" from a boat just behind. Boyes and Trevethick got to shore, safe but shaken. Within two hours the boat was patched—using their fiberglass-repair kit—and the expedition was back on the water.

What's telling from this episode, besides the fast recovery, is how it exemplifies the Okavango Wilderness Project's harvest of data. From the observations gathered that hour, by electronic device and human eyeball, recorded instantly into an elaborate data-vacuuming system, we know that the Cuíto River thereabouts has a strong current, a sandy bottom, and not much aquatic vegetation but harbors smallmouth bream, among other fish. We know that Trevethick made note of a pied kingfisher, then a malachite kingfisher, then a blacksmith lapwing, perched on shoreline limbs. We know the longitude and latitude at which the mishap occurred, to at least 12 decimal points of GPS accuracy. We know that Steve Boyes's pulse rate (as registered by his Suunto watch, also patched to the system) rose abruptly from 81 beats a minute to 208 beats a minute at 10:57 a.m. And we can assume that 208 is the normal heart rate for a healthy young man trying to outswim the watery gallop of a hippopotamus.

WHEN I JOINED Boyes's team on the Cubango River, almost two years later, their data-gathering regimen had advanced to include more categories of information. One morning I watched a young Namibian named Götz Neef assess his overnight catch in a fish trap: a largemouth bream, an electric elephantfish of the sort called a Churchill, a squeaker catfish, and more— weird creatures to me, subtle data points of biogeography to anyone who knows African fishes. Such sampling and collections, analyzed by ichthy- ologists allied with the project, would help reveal how the fish fauna of the Cubango differs from that of the Cuíto and how both may contain unique species or subspecies, distinct from anything else in the region.

Along the Cuíto, for instance, the researchers found what may be a new species of *Clariallabes,* a genus of eellike, air-breathing catfish, this one evidently adapted for wriggling through the saturated peat bogs. Other taxonomic specialists, based in Angola, South Africa, and England, have also assisted the project with field collections—of amphibians, reptiles, insects, small mammals, plants—and the continuing work of identification and analysis. Frogs and dragonflies, with their aquatic immature stages, are sensitive to pollution and can be especially telling as indicators of water quality. One group of peculiar rodents, known as vlei rats in the Afrikaans slang (suggesting that they inhabit transient ponds, or *vleis*) and notable for the smallness of their territories, seems to have diversified into unique species here in the highlands.

"Angola is the missing link," a small-mammal biologist named Peter Taylor, one of the project's experts, told me, "for understanding the pattern of radiation of these beasts." Boyes's goal was to assemble such facts into a mosaic portrait of this two-river system, in its biological and hydrological particulars, to support protecting it for its own sake and the sake of the Okavango Delta.

Besides trapping fish, Neef also saw to the gathering of water-quality data from two delicate sensors as we paddled downstream. And he managed the multifarious photography: one 360-degree camera on a tripod, plus two DSLRs angled out from the bow of his mokoro, snapping frames at five-minute intervals. In the evenings at camp, as darkness fell, Neef deployed a bat detector, a little yellow box that captured high-frequency blurts of chiropterans, used later to identify the species. Other expedition members recorded data about birds, reptiles, human activity. Boyes himself was the principal ornithologist on the rivers, calling out sightings—giant kingfisher, hammerkop, white-fronted bee-eater, lilac-breasted roller—which a young Angolan biologist, Kerllen Costa, entered with GPS tagging into a tablet. Costa's sister, Adjany, an ichthyologist and a National Geographic Emerging Explorer, also assistant director of the project, served as liaison to Angolan officials when she wasn't aboard one of the expedition's mokoros. Boyes's team also included field crew members from Zimbabwe, South Africa, Namibia, the U.S., and of course, Botswana, homeland of his most skilled mokoro boatmen, recruited from the Okavango Delta itself.

Captain of my mokoro was Tumeletso Setlabosha, known to everyone as Water, a small but powerful Wayeyi man who grew up in the central delta. His mother gave him the aqueous nickname because his birth occurred in a pool of water, when she was traveling through the lagoons. Asked his age, Water said he was 54 on land, but "when I'm paddling, I'm 25." He tolerated me in the bow of his boat for a week, as I gawked and scribbled notes and did my best with the paddle.

HUMAN PRESENCE ALONG the upper Cubango was sparse, even though the end of the war had allowed people to return to their villages in these southern hinterlands once controlled by the UNITA rebels. Paddling between reed-lined banks, we saw the occasional beached mokoro, a lonely fishing camp, a cow here, a goat there, a few women washing clothes or making *kashipembe* (moonshine) from jackalberries or other wild fruit in a simple still. Days later, farther downstream, we saw more people, more boats, more livestock, corn crops, a soccer field, a few motorbikes. At night, beyond the trill of crickets, we heard the roar of trucks, and the clatter of their springs taking washboard at full speed, on a bad but important dirt track paralleling the river. That road leads to a border crossing with Namibia, through which supplies can roll in and Angolan timber can roll out. Apart from timber and illegal bush meat and water, the Cubango Valley has little to offer the wider world. No one, it seems, has yet found diamonds or gold or oil in this corner of Angola. Clean water: *That* is the oil and the gold.

One day about noon we beached on the left bank above a small rapid, and because a fully loaded mokoro is too fragile and clumsy for white-water daring, we scouted the line. As we walked, Boyes spotted a hippo snare of stout wire, camouflaged with reeds and placed along a haul-out path used by the animals. There is nothing more piteous than the howl of a hippo in a snare, he said, and clipped through the wire with his utility tool. Boyes had deep sympathy for the needs of people along the Cubango, and he recognized that their progress toward better lives must be part of any arrangement for protecting the two rivers, the water flow, the biological riches of southeastern Angola, and the Okavango Delta. But hippo flesh for meat and hippo teeth to be sold as ivory are contraband commodities that the Cubango can't sustainably surrender.

Another day, we came off the water early to avoid camping near a village called Savate, a place known for land mines still lurking around its perimeter. We beached upstream, at a dirt landing where it was cow pies, not mines, among which we had to step carefully. Children watched us unload an ungainly amount of gear: tents and tables and boxes of food, duffels, folding stools, fancy electronics. Women came with piles of washing to this beach, their regular laundry spot, and had to work around our flotilla of canoes. A tethered donkey grazed nearby. A donkey was wealth. After sunset, by the time our hearty dinner of beans and rice came off the campfire, smelling good, the children had disappeared. I wondered what impression of us they took.

SEEN FROM A Cessna, 500 feet above northern Botswana, the Okavango Delta resembles a mottled carpet of ovals and streaks and patches, gentle rises and swales, a rich pattern textured largely in shades of green and brown. The lagoon waters appear almost black from overhead; the channels and oxbows gleam silver when reflecting a low afternoon sun. At the center of small islands, ringed by trees, lies the whiteness of precipitated salt. Aloft in your little plane and moving slowly, you get a sense of the dynamic heterogeneity below. You see how water has nudged and carved and shaped land over time, opening new channels, closing old ones, rising and falling by season, filling pans, then leaving them to dry, encircling islands, respecting subtle ridges, changing its imperatives and benefices from year to year, and thereby sculpting an extraordinary ecosystem hospitable to fish and crocodiles and long-legged birds and mammals that don't mind having wet feet. That's how I saw the delta, after my time in Angola, thanks to John "Tico" McNutt, a veteran American conservation biologist.

McNutt, a friend of my friend Chris Johns, *National Geographic*'s editor in chief at the time, met me at the small airport in Maun serving Okavango tourism. From there, he flew us to the research camp from which he had worked for nearly three decades, studying the endangered African wild dog. With his breadth of curiosity and involvement, he probably understands the ecological and political dynamics of the delta as well as anyone. Besides showing me dog packs on the ground, he gave me four days of eye-in-the-sky perspective and commentary, even while flying the plane and listening for his collared dogs on the VHF telemetry receiver.

On the left, McNutt said, that's Chief's Island. On the right, the old Mogogelo floodplain, which once carried water almost all the way to his camp. We gazed down at large herds of lechwe, some reedbucks and impalas, termite mounds rising cream-colored at the center of small islands, hippo tracks like claw marks across the floodplain grasses, elephants casting long shadows in late afternoon. "There's no vultures," he noted. "They should be roosting in these palm groves. Should be vultures all over." But vultures are hated by poachers for giving away the positions of fresh elephant carcasses, and are therefore killed by poisoning the left-behind meat. The Okavango, even with the taps open, has its problems.

We flew north across low plains of reeds and papyrus, islands large and small, serpentine channels, until McNutt said: "Somewhere right here would be the fault line, where everything starts to distribute. From the panhandle."

The panhandle is a stretch of slow-moving water, contained by ridges that rise above swampy lowlands, spreading from river to wide sheet just south of the Namibian border and gliding southeast from there toward the line McNutt mentioned, known to geologists as the Gumare Fault. Beyond the fault line lies a sunken, flat trough, partly filled with sediments but still nearly the lowest zone in the Kalahari Basin, across which the Okavango waters open broadly into their flower-blossom shape. The blossom petals come to a dead stop, though, at another pair of diagonal faults, marking the southeastern boundary of the delta. Meeting those natural dams, what remains of the surface water slides westward into a natural impoundment, Lake Ngami, or sinks away into the sands. South of all this: salt pans and desert.

Amid the complexities of water delivery and biological enrichment, from the headwaters to the delta, from Angola through the Caprivi Strip to Botswana, several factors are especially fateful. The delta itself receives rainfall but not much, and mostly during the summer months of December through March. Angola's central highlands receive far more, a great wet bounty, roughly 50 inches annually, which saturates the peat deposits and sands of the upper Cuíto floodplains and then slowly, after delay, flows down the Cuíto and its tributaries. Those rains feed the Cubango too, but the Cubango River catchment lies on steeper, rockier substrate, so the seasonal rainwater comes gushing down fast.

The result of these asynchronies is that the Okavango Delta gets three separate pulses of water annually, giving it a longer and more varied supply of moisture than most freshwater wetlands enjoy. Freshwater coming in pulses, spread across the year, distributed in an ever changing pattern of channels and pans and lagoons, nurturing vegetation of many types, fertilized by the dung of elephants and hippos and impalas—all this is a good recipe for biological fecundity.

The biggest challenge the Okavango Wilderness Project faces is not just to understand this complex system—that's hard enough—but also to persuade Angolan officialdom, and the Angolan people, to preserve the Cuíto and Cubango Rivers roughly as they are, flowing free and clean, without much pollution or diversion, through landscapes mostly undamaged by timber harvest, charcoal-making, forest burning for hunting drives, commercial extraction of bush meat, agricultural schemes demanding high inputs of fertilizer, mining, or other destructive uses. It's an urgent task and not an easy one.

Some optimists propose that landscapes along the Cuíto and Cubango could become international tourism destinations themselves, sites of luxury lodges drawing visitors to see restored populations of wildlife, such as the Angolan giant sable (a magnificent black antelope, not to be confused with the Russian sable, a cousin of mink), that were mostly lost during decades of war. Maybe such attractions could be included in a regional circuit, they suggest, along with more famous camps in the Okavango. Another hope is that the Botswana government and its tourism industry might recognize the jeopardy of their wonderland—recognize that without the Cuíto and the Cubango, there is no Okavango Delta—and act with foresight, offering a compact of payments to Angola for continued delivery of the water. Call it ransom or call it a "water bond" (as Steve Boyes does), it seems rational. Rationality and foresight might be improbable expectations when it comes to intergovernmental relations over resource issues, but the Okavango Delta itself is an improbable phenomenon deserving exceptional concern, imagination, and effort.

Meanwhile the changes in Angola, as Boyes told me, were happening fast. "If we started this work in three years' time, there'd be nothing left to protect." The future is coming like a river that flows through other people's lives.

....

It's been five years, not just three, since Steve Boyes made that remark, during which he and his colleagues have been busy and productive. When I reached him recently, he was preparing to leave on another exploratory river trip, this one for two months down the Lungwevungu River, "the real source of the Zambezi deep in the Angolan Highlands," as he describes it. Since the trip I joined in 2017, he and his team have completed nine more ambitious land and river expeditions covering various parts of what he calls the Okavango-Zambezi Water Tower, that elevated and extremely rain-soaked portion of the Angola Plateau draining southeast through multiple tributaries to those two big rivers, the Okavango and the Zambezi. And in December 2018, his Okavango Wilderness Project (OWP) signed an agreement with the Angolan government, stipulating that OWP will work with local Angolan communities on "conservation economy" enterprises benefiting the communities, and on preserving their natural heritage, while protecting the quality and flow of the water and the other ecosystem services. It's the right combination of goals.

Down in the Okavango Delta, meanwhile, OWP is working to establish "traditional reserves" managed by local communities, roughly on the lines of "community conservancies" as they exist in Kenya, through which such communities can benefit directly from tourism and other income streams as they protect the landscape, waters, and wildlife on their own lands. With support from corporate sponsors as well as the National Geographic Society, OWP has met its ambitious budget needs for the next five-year phase, and in 2019, the OWP team jointly received the Rolex National Geographic Explorer of the Year award. Steve Boyes, fighting through challenges of varied sorts, seems as ebullient as ever.

....

PEOPLE'S PARK

·❦·❦·❦·

Conservation and Human Rights
in Gorongosa

At a gathering of conservationists some years ago, of which I remember no other details, I saw a man named Greg Carr present a talk on Gorongosa National Park, in Mozambique, and the ongoing effort to restore its ecological health, species diversity, and security after its devastation during a long civil war. He spoke also of measures being taken to make it a national park that added to, rather than subtracting from, the life quality of the Mozambican people who live nearby. I was struck by Carr's unpretentiousness, his sensibleness, his slightly disheveled and boyish manner, and his passion for this effort. I was unaware, back then, that this man happened to be the wizard tech entrepreneur and philanthropist, maybe not quite a billionaire but close, who was putting truckloads of his own money, as well as his time, energy, credibility, and love, into the restoration and reimagining of Gorongosa.

Several years later, the editors of National Geographic *suggested to me, before I could suggest to them, that Gorongosa merited a story. For that assignment, I had the chance to spend considerable time with Greg Carr, who served as a sort of personal guide and generous host, as well as with members of the team he had assembled at Gorongosa. Carr's primary message in our dealings, implicit and explicit, was always: For god's sake, don't write about me, write about them. By "them," he meant the people he works with, the people of greater Gorongosa, the staff and citizenry he loves and serves. Wildlife in Gorongosa National Park, decimated by years of civil war, is rebounding. Carr understands that the animals' future depends on providing hope for the people who live nearby.*

Again, I made two trips to research the story. My editorial links at National Geographic *had changed by this time, Susan Goldberg having succeeded Chris Johns as editor in chief, and Jamie Shreeve having taken over from Ollie Payne as my story editor. Jamie became, as Ollie remained, not just an editor but also a friend. Both he and Susan were keen and supportive professionals, tossing me into the sorts of briar patches I enjoyed, and sometimes pushing me hard to make a story better or more engaging. I didn't always agree with an edit, but I always listened and tried to give what was wanted. That's basic to the writing profession in general (including the production of books) and to the magazine-writing craft in particular, especially as practiced for* National Geographic, *with its combined emphasis on photography, graphics, and text: You remember that you're one player on a team.*

After my first trip to Gorongosa, I delivered to Jamie a draft that carefully covered most of the sociocultural aspects of the Gorongosa project—education for village girls, agricultural development for people in the buffer zone around the park, the scars left by the civil war, the vision of "peace parks" that came from Nelson Mandela. Yes, those are good and important, Jamie told me after reading it— and then came the dangler that all writers dread hearing from an editor: "but . . ." But Susan and I miss something here, he said: big animals and drama. Could you go back to Gorongosa in time for the annual elephant-collaring-by-helicopter work? We'd like some great beasts and action as well as the sociology. Sure, I said.

Back in the park, along with my photo partner on this assignment, the serious but never-too-serious Charlie Hamilton James, I saw more elephant-and-helicopter drama than I could put into the article. At one point, for instance, our chopper hovered too long above the tall, dry grass and set fire to it, a sizable conflagration, leaving Charlie and me and some others to flail at the flames with jackets and palm leaves while biologists worked fast on a tranquilized elephant, knocked cold and lying helpless in the path of the spreading blaze. Oy, it could have gotten very ugly. It didn't, not quite. But that's a different tale for a different time. Here I'll merely assure you that no humans and no pachyderms were injured in the making of this story, and my jacket was an old one anyway.

. . . .

Pedro Muagura, a burly man with a soothing smile, gentle in manner and joyous, is chief warden of Gorongosa National Park, in Mozambique. One day there, he told me a somber story: of how his uncle Justin was beaten half to death for the crime of eating an orange.

That was back in colonial times, before independence from Portugal in 1975, when Black Mozambicans worked in the white-owned orchards—apple, peach, lychee, orange—but were forbidden to own such trees themselves, or to eat their fruits. Mozambique, on the southeast coast of Africa (as well as Angola on the southwest coast and Portuguese Guinea, to the northwest), had been declared part of the greater Portuguese state. But unless you had a card saying *assimilado,* "assimilated," certifying your surrender of cultural identity and pride, the fruit prohibition was just one among many onerous constraints upon you as a Black Mozambican. Another: no schooling beyond the fourth grade. Muagura's uncle Justin was not assimilado. He had the temerity to eat an orange—and more provocative still, he tried to plant an orange tree—for which he was punished brutally by white farmers who controlled the orchards. Bones broken, crippled, uneducated, unable to work, fed and carried about by his wife, he lived another seven years and then died young.

"He was beaten before independence"—just before, in 1974—"and he died after independence," said his nephew Pedro, as we sat on the porch of my bungalow at the Gorongosa headquarters camp. "The system at that time was very bad."

What came after independence in 1975 was bad too: 16 years of ruinous civil war, fought mainly between two factions contesting left-versus-right ideologies but also the ownership of power, and its perquisites, in a political vacuum following the hasty exit of the Portuguese. On the left was FRELIMO (its name an acronym for the Mozambique Liberation Front), which took power early and held it, with support from the Soviet Union and Cuba; on the right was RENAMO (the Mozambican National Resistance), a guerrilla army that lurked in the boondocks, battling and raiding sporadically, with backing from white-ruled South Africa and the United States. Straddling between them were the ordinary people of Mozambique, roughly a million of whom died in the fighting or from starvation, while much of the country's infrastructure, institutional framework, and agriculture were destroyed. By 1992, when a cease-fire took hold, Mozambique was among the most woebegone countries in Africa.

Pedro Muagura was young enough to have missed the colonial tyrannies, survived the civil war, and gotten educated, first in Mozambican schools, then during a scholarship in Finland; from there he went to Tanzania, and came back to Mozambique for degrees in wildlife management and forestry. He worked as a lecturer at an agricultural institute in the town of Chimoio, 60 miles west of Gorongosa, until his recruitment in 2009 to what was then called the Gorongosa Restoration Project, an effort to revivify this great national park that had been devastated during the civil war. A decade later, as chief warden, he could look out across the recovering landscape and wildlife populations, and he could eat an orange if he wished, and he could smile.

ON A WARM morning at the end of the dry season, early November, a red-and-black Bell Jet Ranger helicopter raced eastward above the palm savanna of Gorongosa.

Mike Pingo, a veteran pilot originally from Zimbabwe, controlled the stick; Louis van Wyk, a wildlife-capture specialist from South Africa, dangled halfway out the right rear side holding a long-muzzle gun loaded with a drug-filled dart. Seated beside Pingo was Dominique Gonçalves, a young Mozambican ecologist who served as elephant manager for the park.

More than 650 elephants inhabited Gorongosa at this time—a robust increase since the days of the country's civil war, when most of the park's elephants were butchered for meat, and for ivory to buy guns and ammunition. With the population rebounding, Gonçalves wanted a GPS collar on one mature female within each matriarchal group.

She picked a target animal from a group running amid closely spaced palms, and Pingo took the helicopter in as low as the trees permitted. Ten elephants—adult females, small calves at their sides, subadults also staying close—fled the throbbing din of rotors. Van Wyk, forced to make a longer shot than usual, nevertheless put his dart into the chosen female's right buttock. I watched all this, along with my photographic partner, Charlie Hamilton James, from a second helicopter, chasing just behind.

Pingo landed, we did too, and when van Wyk and Gonçalves jumped out, we followed, clambering through trampled grass toward the sedated elephant. Moments later a ground team arrived with heavier supplies, technical helpers, and an armed ranger. Gonçalves placed a small stick in the end of the elephant's trunk, propping it open for unimpeded breath. The animal, sprawled on her right side, began snoring loudly. One technician drew a blood sample from a vein in her left ear. Another helped van Wyk scooch the collar under the elephant's neck.

Gonçalves, wearing medical gloves, took a swab of saliva from the animal's mouth and a rectal swab from the rear, sealing both into vials. She pulled a long plastic sleeve onto her left arm and reached deep up the elephant's rectum, bringing out a handful of fibrous, ocher poop that would be used to analyze the elephant's diet. The elephant's great flank heaved up and down gently in rhythm with the trombonic susurrus from her trunk.

"Louie, can you tell if she's pregnant?" Gonçalves asked.

"She's due soon," he said, noting the watery milk leaking from the elephant's distended breasts.

The growth of the elephant population was only part of the encouraging news from Gorongosa. Most of the big fauna, including lions, African buffalo, hippos, and wildebeests, were vastly more numerous now than in 1994, shortly after the war. In the realm of conservation, where too many indicators herald gloom and despair, success on such a large scale is rare.

Van Wyk finished fitting the collar and Gonçalves packed up her samples. Van Wyk injected a wake-up drug into an ear vein, and the crew backed off to a safe distance. After a minute, the elephant stood, gave her head a groggy shake, and strode away to rejoin her group. Tracking data from the collar would tell Gonçalves and her colleagues how the elephants were moving across the landscape, and would alert them when the group was crossing a park boundary toward a farmer's field, so the farmer could take steps to save the crops. Elephant management encompasses studying them, keeping them safe, and preventing situations in which they might make themselves unpopular.

This is how it's done, anyway, by the Gorongosa Restoration Project, a partnership launched in 2004 between the Mozambican government and the U.S.-based Gregory C. Carr Foundation. For elephants and hippos and lions to thrive within a park boundary, you need to ensure that the humans who live outside the boundary thrive too.

STRETCHING ACROSS A floodplain at the south end of Africa's Great Rift Valley, encompassing savanna, woodlands, wetlands, and a wide pan of water called Lake Urema, Gorongosa was once a hunting reserve: Portuguese colonial administrators established it in 1921 for their sporting pleasure by removing the people who once shared the landscape with wildlife. In 1960, when first designated a national park, it harbored about 2,000 elephants, 200 lions, and 14,000 African buffalo, as well as hippos, impalas, zebras, wildebeests, eland, and other iconic African fauna.

But its remoteness became its undoing. In the ruinous 16-year civil war that followed independence in 1975, Gorongosa served as a refuge for the right-wing RENAMO, rebel forces who received military support from neighboring Rhodesia (now Zimbabwe) and South Africa. When government troops came to challenge them, there was fighting on the ground, rocket shelling of the park headquarters, carnage across the savanna. In addition to the elephant slaughter, thousands of zebras and other big animals were killed for food or trigger-happy amusement. A cease-fire halted the war in 1992, but poaching by professional hunters continued, and people in surrounding communities set traps for whatever edible animals remained. By the turn of the century, Gorongosa National Park had been wrecked.

Circumstances were just as grim on the lands surrounding the park. About 100,000 people lived in what restoration planners would later call the buffer zone—mostly families growing corn and other subsistence crops, barely able to feed themselves, their children shorted on education and health care.

When the soil tired and the corn failed to thrive, the farmers would cut forest, burn the slash, and try again on a new patch. Eventually their cutting and planting expanded from the lower slopes of Mount Gorongosa—a granite massif that looms 6,112 feet above the western boundary of the park—to the higher, wetter zones. Once topped by thick rainforest, the mountain is the source for the Vunduzi River, which carries water to the park and its rich floodplain. By the start of the 21st century, large swaths of forest on the mountain and elsewhere throughout the 2,000-square-mile buffer zone had been stripped away.

The beginning of the end to this cycle of desperation and loss came in 2004, when the president of Mozambique, Joaquim Chissano, visited Harvard University for a lecture at the invitation of a wealthy and curious American named Greg Carr. In 1986 Carr and a friend had created a company called Boston Technology, which presciently offered ways to connect telephone systems with computers. Another successful enterprise followed, and by 1998, not yet 40, Carr found himself on the receiving end of an $800 million deal. "My hobby was to read paperbacks that I could buy for five bucks," he told me during a conversation at Gorongosa. "It was more money than I needed."

He established the Carr Foundation, a philanthropic entity, before he knew for certain what its purpose would be. Around the same time, he started reading intensely in two distinct realms. The first was conservation, which caught his attention by way of Edward O. Wilson's book *Biophilia*, a meditation on humanity's instinctive bond with the natural world. "That started me reading everything of Ed Wilson's I could get my hands on." Wilson made his scientific reputation as perhaps the world's leading expert on ants, but his body of published work is eclectic and provocative, ranging from technical studies of ant taxonomy and behavior to the Pulitzer

Prize–winning *On Human Nature,* to passionate disquisitions on biological diversity and warnings of its decline, such as *The Diversity of Life.* These awakened Carr as a conservationist. Simultaneously, he immersed himself in the literature of human rights, which took him backward in time from Nelson Mandela to Eleanor Roosevelt and the "Universal Declaration of Human Rights," which Mrs. Roosevelt helped draft for the United Nations in 1948, then to the ideas of Mencius, an early Chinese philosopher, and the plays of Euripides, least conventional and most egalitarian of the great tragedians of classical Athens. Carr followed this line as a sort of two-person study project with his friend Samantha Power, a former war correspondent and then scholar who would later serve under President Obama as U.S. ambassador to the United Nations (and still later, under President Biden as head of the U.S. Agency for International Development). Also working with Power, and putting a hunk of his money to use, Carr created the Carr Center for Human Rights Policy at the Harvard Kennedy School of Government.

Then the two lines converged, when Carr learned that Mandela himself, by then president of South Africa, and Mandela's friend President Joaquim Chissano of Mozambique, had begun collaborating to create what they called "peace parks"—transboundary national parks for the protection of wildlife and the benefit of local people on the lands near them. The two lines of interest became a single, stout rope.

In 2004, Carr met Chissano during the Mozambican president's visit to Harvard. "President Chissano loved national parks," Carr said, "and he invites me to restore Gorongosa." That sounds simple, man-to-man, but the government's deliberations were complicated, as I learned in Maputo, Mozambique's capital, from Dr. Bartolomeu Soto, the former deputy minister of tourism who negotiated the initial management agreement with Carr.

"It was our first public-private partnership for managing parks," Soto told me. "So people were suspicious."

Soto had reassured his colleagues: "We're not *giving* him the park." If it didn't work out during a provisional period, Mr. Carr could leave. "But he's not taking the park *with* him." The government and Carr signed a memo of understanding, after which the Carr Foundation promptly invested more than $10 million toward reconstruction of facilities and reintroduction of African buffalo and wildebeests. But officials still viewed the park (even in its state of dereliction) jealously, and a long-term agreement was more difficult. When Soto presented that to the Council of Ministers, he told me, the minister of finance was skeptical.

"He'll manage the park," said the minister, ruminating. "Do we pay him?"

"No, he pays us," Soto explained.

"What does he get?"

"Satisfaction."

"Only?"

"Yes. Because he's doing biodiversity."

In December 2007, Carr signed this long-term agreement between the Mozambique government and the Carr Foundation, after which the Gorongosa Restoration Project brought further ideas, skills, partners, and funds to the effort—doing biodiversity, yes, but also much more.

THE BIODIVERSITY ASPECT is what's paramount to Piotr Naskrecki, a Polish-born entomologist with one office at Harvard and another at the Edward O. Wilson Laboratory at Gorongosa, named for the man who mentored Naskrecki and inspired Greg Carr. "The social work is tremendous," Naskrecki told me. "But we ended up here because of the biodiversity." He first visited Gorongosa in 2012 on an expedition, led by Wilson, to gauge the diversity of insect life, which they found to be extraordinary. By the time I met Naskrecki, besides running the Wilson Lab, and its master's degree program in conservation biology for Mozambican students, which had 12 participants, he led the effort to create a comprehensive inventory of all life-forms within the park. That inventory, almost impossibly ambitious given the variety of landscapes and habitats comprising Gorongosa, would be nonetheless crucial. "The underlying principle," he said, "is that you cannot effectively protect something you don't know exists." So several times each year he spearheads a team, including experts in this group of creatures and that—insects, reptiles, small mammals, big mammals, plants—who go on multiday blitzes to identify everything they can find. The findings from those surveys, plus the identifications of known creatures and the description of new species, are being used to assemble a comprehensive Map of Life for the ecosystem. A corollary benefit is that such surveys offer great opportunity for teaching Mozambican students the methods of field biology and identification.

Carr and I joined one of these blitzes at a field camp up north, within a former hunting concession called Coutada 12, recently added to Gorongosa National Park. Piotr Naskrecki met us at an old hunting lodge, lately being repurposed but on the walls of which still hung antique rifles and fading photos of proud hunters with their kills, and he drove us by Land Cruiser to the field site, deep within miombo forest. (The word "miombo" is Swahili for a group of related trees, known to science as the genus *Brachystegia,* characteristic of landscapes in south-central Africa that are generally a mix of savanna, grasslands, low forest, and shrublands, found from Angola in the west across to Tanzania and Mozambique.) In the hours before sunset, beyond the perimeter of our tents, everyone got busy—Naskrecki in preparing a sonic experiment to identify katydids (his specialty) by their various songs; Jen Guyton, an American ecologist, setting up a mist net to catch bats, with help from two Mozambican students; Dominique Gonçalves, the young Mozambican who manages the park's elephant ecology project, looking for tracks; another Mozambican student, along with two middle-aged Polish entomologists, deploying a large white screen with a strong light behind it, for attracting moths and other nocturnal insects. After a camp dinner of spaghetti, the night work began, and by 9:00 p.m. Jen Guyton had captured half a dozen bats, including two specimens of Decken's horseshoe bat, a little-known East African species. Her special target was another rare species, Dobson's pipistrelle, not seen in southern Africa for 30 years until she caught one specimen during a similar survey a year earlier. Tonight, it was Decken's but no Dobson's in the net, and the quest would continue in following days. If it were easy, it would be called shopping, not chiropterology.

Next morning, we bushwhacked through the forest, following Guyton and her GPS toward a sunny forest gap where, on a previous visit, she saw some strange bees—massive, black-and-white creatures, like the giant pandas of Hymenoptera—gathering nectar from a flowering vine. Along the way, with Naskrecki as our all-purpose nature guide, we admired aardvark burrows, orb-weaving spiders, massive dung beetles, elaborate fungi, termites, millipedes, an old poacher's camp littered with a gin bottle beside the campfire ashes, and some tiny ants so subtle and obscure that only Ed Wilson could identify them—which, said Naskrecki, on a previous outing Wilson did. Naskrecki told the tale, of Wilson breaking open a small twig that others had ignored and, within its hollow stem, finding these ants. How did you do that? asked Naskrecki, then he realized his question was silly. Wilson only smiled modestly, leaving the message unspoken: *I may be 85 years old, but I'm still Edward O. Wilson.*

At the forest gap, Naskrecki captured two specimens of the mammoth black-and-white bee, which he said was a new species for park records, but probably not new to science. We also sighted another strange bee, even larger, fuzzy and orange, working blossoms on the highest stretch of vine, and so elusive that not even Naskrecki could net it. This was a bee you watch through binoculars. But Naskrecki vowed to come back and catch one. How? "I'll wait."

Returning toward camp, we found something still stranger. A column of driver ants had swarmed upon some poor creature—it looked like a worm, maybe a night crawler—and were killing it with a thousand bites. "I think it's an *Amphisbaena,*" Naskrecki exclaimed. "Holy crap." He had told us over breakfast about that weird, legless form of burrowing lizard, named for the corpse-eating serpent in Greek mythology with a head at each end. Now he rescued this unfortunate victim, brushing away ants, rinsing them off with water, but it was too late. "He's done for." And then, seeing it clearly, Naskrecki revised his identification: It was *Zygaspis,* not *Amphisbaena.* A purple round-headed worm lizard, to you and me. "This is a genus we haven't found in Gorongosa. I have no idea of the species." But in any case, a new point of data in the bounteous index of one park's biological diversity. As the man said: Holy crap.

ON A WET Thursday morning in April, nine little girls jumped rope beneath a sheltering tree in Mecombezi Ponte, a village about 20 miles from the park. They wore dark blue T-shirts with "Rapariga do Clube" (Club Girl) emblazoned on the back and a small round seal saying "Parque Nacional da Gorongosa" on the front. In a semicircle around the girls stood 10 *madrinhas,* or volunteer "godmothers," giving their time and quiet vigilance to help protect these young girls from the jeopardies they faced: forced early marriage, frequent pregnancies, bad health, and truncated education.

The Girls' Club of Mecombezi Ponte was one of 50 such clubs (with more to come) organized and sponsored by the park to augment daily school sessions for some 2,000 girls throughout the buffer zone. On Mondays, Wednesdays, and Fridays, the focus is on literacy. Tuesday's agenda is health and reproduction. Thursdays, as Carr and I saw, are devoted to play. The women clapped and sang while the girls gleefully took turns in the twirling rope. Carr, sporting a T-shirt, shorts, and a two-day growth of beard, joined the line of girls and gamely tried to jump rope. The girls were better.

Carr regarded the Girls' Clubs as a critical part of the Gorongosa National Park resurrection. Deterring men from hunting the park's wildlife—through alternative livelihoods as well as ranger enforcement—is important but insufficient. Women are the fulcrums. If the human population in the buffer zone continues to grow unabated, by way of early marriage of girls and large families, no effort within the park boundaries will be sufficient to protect its landscape and fauna. "But if girls are in school and women have opportunities," Carr said, "then they will have two-child families." It's not an imposed solution. It's part of a phenomenon resulting from women's empowerment. "This is where human development and conservation merge," he added. "Rights for women and children, poverty alleviation, is what Africa needs to save its national parks."

Before departing, we witnessed a small ceremony. A sixth grader named

Helena Francisco Tequesse stepped forward and, from a laminated card, read a declaration of 10 rights and 10 duties of children. "Children have the right to be fed and a duty not to waste food," she read. "Children have the right to live in a healthy environment and a duty to care for the environment."

"This is really exciting," Carr said. "When I came here, the percentage of women in the buffer zone who could read—zero." He asked the girls to say what they wanted to be when they grew up. Each stepped into the dirt circle, said her name, and answered with poise: a nurse, a midwife, a teacher, another nurse, a police officer. By now, with the rain finished and the morning turned sunny, the group had grown to about 30 girls and madrinhas. As we left, they resumed clapping and singing and dancing.

WHAT DO THE education and welfare of young village girls, and the problems of earning a livelihood faced by their parents on a rural landscape, have to do with protecting elephants, bats, and legless lizards inside a national park? Greg Carr's answer: *everything*.

But the connection is indirect, and making it requires vision and patience. Pedro Muagura told me another story during our private chat that reflected the learning curve, even at Gorongosa. It was a meeting of Gorongosa leadership, during his first year with the project, at which he suggested the park should grow coffee on Mount Gorongosa. Most of those present remained quiet, evidently dubious. "But there was a lady who was director of scientific services," he told me, recalling the scene with a mild but enduring sense of grievance. "She said, 'No, you don't understand what is conservation.' I said, 'I do understand, but if you'll give me a chance to explain . . .'" He didn't get that chance, not then. He was the new guy, just a forestry manager; he didn't know conservation from agronomy, and she was adamant. "No, nothing to do with coffee in the park." She was now gone; another scientist had taken over that division, Muagura was chief warden, and the park had established a coffee program in the buffer zone, helping farmers on Mount Gorongosa produce high-value, shade-grown beans beneath canopies of replanted native trees.

Though it lies outside the park's original boundary, Mount Gorongosa is an indispensable part of the ecosystem. The mountain not only captures rainfall and delivers it to the park's floodplain, but it also adds a diversity of altitude, climate, soil, vegetation, and wildlife to the greater Gorongosa whole. In 1969 a South African ecologist named Ken Tinley proposed that the mountain, as well as the plateau and coastal habitats stretching eastward from the park border, also richly various, be combined into a single integrated management area.

Tinley's idea has taken hold as the "mountain to mangroves" vision of Gorongosa. In 2010 the highlands of Mount Gorongosa (the portion above about 3,000 feet) became part of the park. That mountaintop encompasses the source of the Vunduzi River as well as some remote forest (still held by rebels when I visited, despite a recent cease-fire), but across the lower elevations local people continued cutting, burning, and farming. They had little choice. And Muagura's coffee idea, despite a flare-up of the war in 2014–16, when government forces advanced up the mountain to attack the rebel hideout, was blooming nicely. Early one morning, with a small party of people including Quentin Haaroff, a Zimbabwean agronomist and the park's coffee expert, I went to see it.

From a town called Vila Gorongosa, along a highway that marks the west edge of the buffer zone, we ascended by Jeep on a steep two-track that climbs the massif's southern slope, passing fields of sorghum and corn, a few houses and huts, a patch of pineapples. Big hardwood trees, felled by RENAMO soldiers to block the road and thwart government vehicles, had been pulled aside and left rotting. Along the way, I learned that Haaroff had farmed coffee himself, in Zimbabwe—until the day, he told me, when President Robert Mugabe made white farmers unwelcome, and he left at the point of a Kalashnikov rifle. So he knows well that war and farming don't easily mix. Slightly higher, we reached the hospitable elevation for coffee.

"This mountain has got a fantastic environment," Haaroff said. Good humidity, temperatures are cool and don't fluctuate greatly, and there's no frost. "You try to do this in Zimbabwe, and your coffee would be dead by now."

Growing coffee beans and restoring forest in an on-again, off-again war zone is still daunting. But the local farmers embrace the enterprise, as evidenced by the women who came out at night and watered the young coffee plants even during the renewed fighting in 2014. Those plants survived and now flourish, along with many more.

We parked the Jeep and proceeded by foot, crossing a small river on stepping-stones, and inspecting a tree-shaded nursery of 260,000 coffee starts, each one growing from a scoop of soil in a pot-like plastic sleeve. Farther upslope, we moved amid producing trees, bush size and healthy, planted in cross-slope rows and shaded by acacias and other trees. The park currently employed 180 people on this work, Haaroff explained, as a demonstration project. The plan was to show how it's done—coffee plants, shaded by native trees, mulched with compost, weeded by hand, with vegetables, fruits, and legumes as secondary crops between the rows—and then to supply training, tools, coffee starts, and seeds, and to offer a good price for the harvested coffee, bought by Produtos Naturais, a natural-products enterprise within the park's sustainable finance division.

Produtos Naturais processes the coffee at its factory nearby and markets the roasted beans to Mozambican wholesalers and internationally. All profits to Produtos Naturais go to support the people, wildlife, and ecosystem of Gorongosa. The coffee, and other premium cash crops (such as honey and cashews), give local people better livelihoods and wean farmers away from slash-and-burn corn, thereby not just protecting what's left of the mountain forest but also reforesting areas that have been cut. "I'm not a scientist," said Haaroff, "but the birds have come back; the bees have come back. You can just see nature breathing a sigh of relief."

NATURE IS RESILIENT, but its sighs of relief, its trends of recovery and resurgence, require more than reforestation of mountainsides and protections against poaching. A pack of African wild dogs (native predators, also known as painted wolves, and lost during the war) was released into the park in 2018, after weeks of acclimation in a large pen. A small herd of zebras also trotted cautiously from their corral and into the wild. A solitary leopard was spotted.

Black rhinos once roamed Gorongosa as well, but that difficult reintroduction challenge, with high risks of attracting commercial poachers, will have to wait. Full recovery takes time and space. The time dimension is recognized in a long-term agreement between Carr's group and the Mozambican government, renewed in 2018 for 25 years. Of course, even 25 years is just a beginning in ecological terms.

The significance of space—bigger protected areas generally embrace more diversity and greater ecological wholeness—helps explain why Carr and his colleagues, including partners within the government, favor further enlarging Gorongosa in line with Ken Tinley's early mountain-to-mangroves model. They envision a greater Gorongosa ecosystem, all of it protected or sustainably managed, encompassing successful farmers and other local enterprises, and connecting Mount Gorongosa in the west, the park in the southern Rift Valley, large blocks of hardwood forest on the Cheringoma Plateau just east of the valley, and the unique coastal woodlands and swamps on the south side of the Zambezi River Delta. The coastal piece of that puzzle already enjoys some protection as Marromeu National Reserve, a soggy and roadless wilderness rich with African buffalo and birds.

For a panoramic glimpse of all this, Carr and I lifted off one morning in the Jet Ranger with Marc Stalmans, director of the park's science department, and headed east toward Marromeu, passing low over savanna, then palm forest, then the thicker forest of the plateau. Flying over this landscape in 50 years, Carr said, hopefully, determinedly, Dominique Gonçalves or someone else of her generation would see wildlife in huge round numbers: 10,000 elephants, a thousand lions. As for buffalo, maybe 50,000.

"Difficult but doable," Carr added. "I like the idea that it's just on the edge of possible."

"Difficult" is an understatement. The latest aerial count of wildlife in the park, as of October 2018, revealed continuing increases for many species: buffalo up, kudu up, impala way up. In addition to the reintroduction of African wild dogs, populations of zebras, wildebeests, and eland had grown. Patrol sweeps by rangers—261 of them, including a small but growing number of women—were keeping losses to poaching at a minimum. The counts showed that Carr's goals were still a long way off, but if the edge of the possible can ever be realized, it will be here, in Gorongosa National Park.

Mike Pingo lowered the helicopter onto the beach at Marromeu, and during a brief stop there, he and Stalmans and I talked about African buffalo while Carr wandered off. Buffalo need grass, water, and occasionally shade, Stalmans said, but not much else. Before the civil war, there were 55,000 here in the Marromeu National Reserve. After the war, just 2,000. And those 2,000 buffalo survived only because the soggy coastal terrain made them so hard to hunt.

By this time, we noticed that Carr had ditched his shoes and waded far out into the surf, nudging at limits, as he often does, like a little kid. Returning, he started to imagine a beach lodge into existence, right at this site, bringing tourists to enjoy the coast and wildlife, plus a marine research station named after some heroic Mozambican, all together anchoring the great sweep of variegated ecosystem: the mountain, valley, lake, plateau, coastal wetlands, mangroves, beach. "Put it together," he said, "and you've got something extraordinary."

We climbed back into the helicopter and, whirling away, passed above a sizable herd of buffalo, dark and sleek and each with a couple of egrets, blazing white, perched on its back. The birds rose up and away, spooked by our noise, like a flock of guardian angels returning to base. About 500 buffalo, reckoned Stalmans. Five hundred? Stalmans has done a lot of aerial wildlife counts, but still I wondered how he got that estimate so fast. "It's easy," he said drily. "You count the number of egrets and divide by two."

ON ANOTHER DAY, the helicopter carried us northeast to a site along the escarpment of the Rift, with its limestone cliffs, karst caves, and deep little side canyons. We landed near an exposed bank of sandstone, five to seven million years old, where paleontologists from the University of Oxford, at the invitation of the park, had begun looking for evidence of early hominids. That was fascinating enough, but the work hadn't yet yielded dramatic finds, and what sticks in my memory more vividly is a stop we made later

THE HEARTBEAT OF THE WILD

that afternoon, near a small waterfall tumbling off the escarpment. Carr had mentioned this place repeatedly for a week. He wanted me, before I left Gorongosa, to see the climbing fish.

The cliffs were chalky white limestone because this part of Mozambique was once a seafloor. The waterfall in question was a modest one, about nine feet high, with a pretty pool beneath. We waded the stream and approached the lip of the cliff carefully, to gaze down at the pool. *There*, someone said. The fish. Leaping onto the wall. Where? *There, there,* can you see them? I couldn't. They were tiny, minnow-size things, but jumping like salmon, falling back in small splashes, and I had no idea what to look for. I stood there, leaning and squinting. Before I knew it, Carr was halfway down the rock face, climbing toward the pool.

"What could go wrong?" he said, the boy thing again, tempting fate.

But he got down without breaking an ankle. He crossed the pool, chest-deep in green water, and showed me the fish. Here, these, *see?* Then he peeled off his shirt and started seining them. Catching a few, or more than a few, he climbed out the far side, still holding his shirtload of water and fish, and got back to the helicopter. Marc Stalmans, unsurprised by this sort of antic, helped Carr put them into zip-seal bags.

"If I have 12," Carr said, "we'll give one to each of the master's students and have them identify them."

No matter that an expert had already identified these fish—as redeye labeo, related to carp, known for employing their mouths and pectoral fins to climb wet rock. The exercise would be valuable, nonetheless, for the Mozambican grad students. They might use field guides or taxonomic keys. They might run the DNA. It was a good notion, much in character for Greg Carr. But no more in character than his jaunty remark on the rock: "What could go wrong?"

Many things can go wrong, he knew well, but the way forward is to proceed boldly as though they will not.

....

The first thing that went wrong was Cyclone Idai, which hit the central coast of Mozambique on March 14, 2019, not long after my second visit, and swept into the country on a path that smacked Gorongosa National Park with full force. It was one of the fiercest tropical storms ever recorded as coming out of the Indian Ocean against Africa. It lingered for days and caused a major humanitarian crisis in Mozambique, Zimbabwe, and Malawi, killing about 1,300 people and affecting three million. At Gorongosa, even before the big international aid agencies had organized relief, park helicopters and park rangers began delivering food and other necessities to communities that had been isolated and devastated. That emergency effort was consistent with the declared purpose of the Gorongosa Restoration Project: to help support the people as well as the wildlife within a great ecosystem.

The storm slid back out to sea after a few punishing days. The winds died, the waters receded, and the conservation and development work resumed, slowly but determinedly. The painted wolves, whose reintroduction to the park I had witnessed in 2018, survived the storm and thrived, increasing in number and expanding their range. Greg Carr recently emailed me a photo he took when a gaggle of them, curious and beautiful, approached his Jeep on one of the park's dirt roads. He added the cheerful news that there were now five packs comprising, as of the latest count, 123 individuals. The elephant population has continued to rebound as well, up from 650 a few years ago to roughly 800. And in 2020, leopards were reintroduced.

Carr offered some equally good news on the human side. "We now work in 89 primary schools with 40,000 children. We are training 600 schoolteachers and have thousands of children in our after-school clubs." They have also expanded the club concept to preschoolers, with more than 200 children aged four and five enrolled in the Little School Club, to help prepare them for primary education.

And the club concept has been expanded still further, I heard from Vasco Galante, the park's director of communications: teachers' clubs to help prepare local educators and offer enhancements to curricula, and peace clubs to help reintegrate RENAMO soldiers from the resistance into contemporary Mozambican communities and economies. More schools are being built, Galante told me, more trees are being planted, and more shade-grown coffee is being harvested and sold.

Take heart that good things are happening, in smart ways, amid this faraway Mozambican landscape that was once a zone of peril, misery, and death for humans and other living creatures. Gorongosa National Park is still a work in progress, but it shows what can be possible.

GIFT OF
THE WILD

∙|∙|∙|∙

The Legacy of Doug Tompkins and
Kristine McDivitt Tompkins

Douglas Rainsford Tompkins was an American businessman and an outdoorsman, co-founder of the companies The North Face and Esprit, who became an important leader and philanthropist in the realm of conservation, most notably in South America. I never met him. He died before I had the opportunity to visit the extraordinary landscapes in Chile and Argentina that he and his wife, Kris (Kristine McDivitt Tompkins), worked to restore, protect, and donate to the people of those two countries. Like anyone who lives in the American West, though, and who mixes with an outdoor-sports crowd and cares about conservation, I had heard a lot about him.

One thing I hadn't realized was the degree to which Tompkins brought not just his intelligence, will, generous spirit, and millions of dollars to the mission of protecting wild landscapes for the enjoyment of people and the conservation of biological diversity, but also a very keen aesthetic, a visual style. When I started visiting the places he and Kris Tompkins established as parks, with the help of many devoted colleagues, I noticed that aesthetic in all aspects of the built environment within those wild places: the visitors centers, trails, lodges, campsites, signage, architecture, interior design, the giant Tompkins-published books full of gorgeous photographs and conservation messages that sit on coffee tables in the lodges and cabins and guesthouses. It's a graceful, clean, slightly rustic look *that his friends and his widow still refer to in Spanish, half-jokingly, with recollections of his obsessive attention to detail, as "estilo Doug—Doug's style."*

It was Doug's style to do things at grand scale, sometimes impetuously, with great confidence, voracious research, and always an appreciation of the beautiful line. Kris Tompkins has a style of her own, different from but complementary to his. Her quiet and resolute manner, plus the resources the couple has donated and committed, plus the team of professionals they have assembled, now carry the enterprise called Tompkins Conservation into the future.

"It was a desperate time. Doug never got over it."

Kristine McDivitt Tompkins sat before a coffee table covered with colorful maps of Chile and Argentina, talking about the controversy in the early 1990s that swirled around a place called Pumalín, in southern Chile. Pumalín was the chastening early experience that showed her and her late husband, the retired businessman and adventurer Doug Tompkins, how hard it could be to convert Yankee dollars and good intentions into landscape protection in South America.

Beyond the coffee table, the maps, the big windows of this handsome stone guesthouse, built like an aerie atop a small hill, stretched a vista of rolling grasslands, tumbling streams, forests of southern beech, and midnight blue lakes: the stern natural glories of Chile's Patagonia National Park, another Tompkins project, some 300 miles south of the one at Pumalín. Patagonia NP comprises more than 750,000 acres, including the Chacabuco Valley, running west from the Andes. Together with Pumalín and six other parks—created or enlarged by Tompkins persistence, in partnership with the Chilean government, and leveraged with Tompkins-donated lands—this network of wild places totals more than 11 million acres. The breadth and diversity are vast, spanning the length of Chile's southern half, from the Valdivian temperate rainforest of Hornopirén to the rocky islands and glaciers of Kawésqar. But to understand the scope of what Kris Tompkins and her husband have done, as well as the obstacles they have faced, it was best to start with Pumalín. She unfolded the maps and told me the story.

In 1991 Doug Tompkins bought a derelict ranch in the Lakes Region of Chile, a country he knew from youthful visits as a vagabond skier and climber in the early 1960s. Later in that decade he and his first wife founded the outdoor equipment company The North Face, sold that business for not much money, and then established the highly successful clothing company Esprit. By the start of the 1990s, quite rich, divorced, and disenchanted with ravenous consumerism, Tompkins had cashed out and walked away from the business world, devoting his life to the robust sports that first brought him south—mountaineering, skiing, kayaking—and to conservation.

His plan to restore the ranch's native vegetation morphed into a bigger idea. He created and endowed a private foundation, the Conservation Land Trust, and through it made purchases to assemble two big blocks of mostly wild land, Pumalín North and Pumalín South. Between them lay another parcel, called Huinay, then owned by the Pontifical Catholic University of Valparaíso, which was willing to sell. But powerful political interests, including then President Eduardo Frei Ruiz-Tagle, opposed the sale. Kris McDivitt entered the picture at that point, having recently retired as CEO of another clothing company, Patagonia, and bringing her own wealth and convictions, which aligned well with those of Doug Tompkins. She and Tompkins were married in 1993.

Kris Tompkins is a small, forceful woman with a clinical intelligence. She reminisces without emoting. Huinay, yes, that was the piece that would have united Pumalín, she told me. It amounted to roughly 130 square miles, not large compared with Pumalín North or South, but belting the Chilean mainland at one of its narrowest points, from the Gulf of Ancud to the Andean summits. Their efforts to buy it aroused suspicion, resistance, rancor. They were taking agricultural land out of production, some people groused, with all this buying and protecting. They were killing jobs. They were shaping "a fiefdom" in Chile.

Such reactions continued throughout the 1990s and into the early years of this century, as the couple expanded their land buying and protection to other parts of Chile (including the Chacabuco Valley, where she and I now sat). Who were these grasping gringos, and what nefarious plans did they have? Were they looking to build a nuclear waste dump, or provide military bases for Argentina, or steal away Chile's water? Or did they just want to turn large chunks of Chile into their own private getaways?

In reality, their goal at Pumalín was to buy land, create a park, and give it to the nation. But Chile had no tradition of private philanthropy outside of church and education projects. Such unfathomable generosity from a pair of North Americans seemed patriarchal at best, sinister at worst. Huinay was especially sensitive because, though smallish, it stretched from border to border. If rich gringos owned that property, critics argued, the country would be cut in half.

"We had four or five years of being despised," Kris Tompkins said. "People thought we were a cult."

THROUGHOUT 21 YEARS of marriage, with their multiple far-flung properties and projects in Chile and Argentina and their restless interest in landscape, the Tompkinses spent considerable time in small, private airplanes. He had 15,000 hours as a pilot. She took the controls often but, never licensed, not for landings or takeoffs. "That's when I'm happiest, flying," she told me. They always thought they would die together, she added, because of all that bouncing around in the Cessna or the Husky amid these Andean canyons and peaks.

It didn't happen. He died of hypothermia on December 8, 2015, at a hospital in the regional capital, Coyhaique, after suffering prolonged immersion in a cold Chilean lake on a disastrously unlucky day, when the winds came up, the waves rose high, and the rudder on his kayak malfunctioned. The boat capsized, and the driving chop kept him and his paddling partner, the renowned climber Rick Ridgeway, from reaching shore. Ridgeway was rescued after an hour and barely survived; Doug Tompkins did not.

Kris Tompkins got the news by phone—a vague version, about an accident and maybe a fatality—then drove six hours to the hospital where her husband had been pronounced dead. "Him leaving so quickly matches whatever that marriage was," she told me. "Grief is just a continuation of whatever the relationship was that you had." Intense lives shared, intense grief. So be it.

Her aviation nickname during their years together, for communicating by radio, was Picaflor, Spanish for "hummingbird." Doug's handle was Águila, meaning "eagle." Between the two of them, more intimately, those transmuted to "Birdie" for her and "Lolo" for him. But if she is birdlike, it's in the manner of a storm petrel, buffeted and doughty, not a hummingbird. For the past few years, she has pursued alone, and only more fervently, the effort they began together.

"It's what kept me from going with Doug," she said. Giving up, she meant, lying on the widow's pyre. "I couldn't imagine life without him."

Instead, she refocused on their goal of leveraging Tompkins landholdings into a grand portfolio of national parks, scattered across Chile and Argentina. That took three years, but the process accelerated quickly after Doug's death. Within two weeks of burying him, she reached an agreement to protect an enormous wetlands ecosystem known as Iberá, in northern Argentina. In Chile, by the end of March 2019, she finalized her commitment with the government to combine a million acres of Tompkins land with 10 million acres of government-held land to create five new national parks and enlarge three others. What once was the private reserve of Pumalín is now a public treasure: Pumalín Douglas Tompkins National Park.

AFTER LUNCH AT the guesthouse, Tompkins took me walking to see some of the local landscape. Behind the main lodge of Patagonia National Park, a service road led to a footpath up a creek drainage. We paused at a very small cemetery, square within a stone-pillared fence, with 10 graves marked by wood crosses and small shrines, plus one vertically planted slab of stone, upon which was carved:

DOUGLAS RAINSFORD TOMPKINS
Birdie & Lolo
03-1943 12-2015

Staff members chose the headstone inscription without consulting her, but Tompkins told me it suits her fine. She is briskly unsentimental in her conversations about her husband and his end, but unsentimental is not unemotional, and sometimes, she said, she comes back to this gravesite and lies on the grass quietly, remembering, communing.

The foot trail snaked out across stony hillsides and grassy flats tufted with neneo bushes, yellow-flowered and spiny, rounded in profile so that from a distance they looked like coral heads. It crossed a creek, shaded by beech trees, then climbed toward a campground, simple but well kept for visitors, and looped back toward park headquarters. At one point, I noticed a small pile of dried, bone-white scat. Yes, puma, Tompkins said, picking up a lump and breaking it open to show me the compacted, undigested fur of whatever prey animal this cat had eaten. The rebound of pumas in the Chacabuco Valley is one dimension of *rewilding*, as they call it, a major goal for Tompkins lands in Chile and Argentina that have lost signature elements of their aboriginal fauna. Rewilding means increased populations of pumas and huemuls (south Andean deer, an endangered species) and Darwin's rheas (a large, flightless bird) here in Patagonia National Park, plus other wildlife recovery efforts and reintroductions elsewhere.

Rewilding is also controversial, especially when it involves the return of predators such as the puma or (up in that great Argentine wetland, Iberá) the jaguar. Only a combination of daring and patience could make it happen, and much of the patience has been Kris Tompkins's.

"Doug was the bomb thrower," I was told by Gil Butler, a friend of theirs and a peer in conservation philanthropy. "Kris is, 'Let's go get it done.'"

ON THE ARGENTINE side, Tompkins rewilding initiatives are proceeding busily at Esteros del Iberá, in the northeastern corner of that country. It's a vast, soggy ecosystem, a paisley mosaic of marshes, dark-water channels and sloughs, lagoons, platforms of floating vegetation, hummocks just high and dry enough to support tiny patches of forest, and some areas of solid savanna. Caimans and waterbirds are abundant, and with luck you can spot a yellow anaconda. Sunshine presents it all brilliantly—the name itself is from the Guaraní language, *y berá,* and means "shining waters."

Iberá lies within Corrientes Province, a mostly rural region bordered by Paraguay, Uruguay, and Brazil, with a strong element of Guaraní culture and language and an ethos of frontier independence. Iberá's history for a century included marginal cattle ranching, as well as hunting for meat and hides; local people often traveled by boat or wading horse, but there wasn't enough terra firma to support many humans or cows. The alternative future was trending toward commercial-scale rice farming and pine plantations.

Then, in 1997, Doug Tompkins happened to visit. He became intrigued with the place and, one summer day, flew his wife back for a look. "We got out of the plane, and I just said, 'Hey, let's get out of here,'" she told me. "It's hot, it's buggy, it's flat as a pancake. Get on the plane." But he saw something she didn't—its biodiversity, its possibilities—and bought a ranch on an island amid this great swamp without even discussing it with her, a rare thing. That ranch, Estancia San Alonso, became the first Tompkins foothold in Iberá and, eventually, because of its remoteness, a logical site to begin the most dramatic act of rewilding: the reintroduction of jaguars.

Not far from the San Alonso ranch house stands a cluster of well-engineered enclosures: stout rebar fencing and steel poles, 16 feet tall, T-shaped at the top to prevent animals from climbing out, electrified wire around the inner perimeters. Jaguars can be restless, especially when caged, and athletic.

Each enclosure also contains a tree platform, low brush, or some other natural furniture to provide cover. Eight jaguars were in residence when I visited, including several adult breeders borrowed from zoos and a pair of year-old cubs, born there and being raised for release. The cubs inhabited a larger pen at the back, with plenty to eat but no human contact—even glimpses of their keepers minimized—so that when liberated, they would fear people, not associate them with food, and would possess other good, wild, survival habits.

I watched as a live capybara—a native rodent, huge and meaty—was introduced to one pen. But the adult female jaguar inside either wasn't paying attention or wasn't hungry. She would find it in due time. A big male jaguar known as Nahuel paced back and forth along a fence line, muscles rippling under his smooth, patterned fur.

These cats are ferocious as well as beautiful, of course, and will kill livestock in any area where cows and sheep have supplanted their natural prey. Isla San Alonso is now cow and sheep free, its grass supporting many marsh deer and an almost comical abundance of capybaras (thanks in part to the long absence of their jaguar predators), some of them topping 150 pounds. That's why San Alonso was the right place to start. The first releases were impending. To reestablish jaguars throughout a wider area of Iberá would be more complicated, requiring both social acceptance and available wild prey.

Tompkins Conservation (as the whole two-country enterprise came to be known) was addressing that with a campaign of education and events, intended to nurture jaguar appreciation as part of the proud legacy of Corrientes Province. At a first birthday party for the two jaguar cubs, in the town of Concepción, I watched more than a hundred people, adults and kids, celebrating in a courtyard amid brightly painted animal murals, guitar and accordion music, small children twirling colored streamers, free cookies in the shape of a jaguar paw, and a puppet show. Kids took turns posing for pictures in front of a huge jaguar poster, each kid delivering a jaguaresque roar. *"Corrientes Ruge*—Corrientes Roars," read the poster legend.

THE REWILDING EFFORT also involves red-and-green macaws, pampas deer (a threatened species), collared peccaries, giant otters, and giant anteaters. Some of the preparatory work with those animals occurs at a quarantine compound, down a narrow side road and behind two layers of fencing, near the town of Corrientes, the provincial capital.

A local woman named Griselda "Guichi" Fernández, who formerly worked as a cook and cleaner, became the expert foster mother to the little orphaned anteaters raised at the compound, each of which had its own pen. On the afternoon I stopped by, Fernández offered a bottle to one, known as Quisco, which clung to her lovingly as his very long snout found the nipple and his noodlelike tongue came out to lap the milk. After the feeding, he luxuriated in the attention as Fernández tickled his tummy, but that easy intimacy couldn't last.

"They are such instinctive animals, they can't be raised as pets," she said. "After they're a year old, they have big claws and they're dangerous."

Such orphans often are left behind when the mother is killed in an altercation with a hunter and dogs, during which a dog sometimes dies too. An adult giant anteater is a magnificent, improbable creature with brindle fur down its back, white chaps, a racing stripe of black, a huge furry tail that can serve as a blanket when it sleeps, a gracefully curved snout that works like a vacuum attachment, a tongue half the length of its body, and those claws. Eight adults resided in larger pens not far from Quisco's, and when Fernández arrived with their dinner—a slurry of cat food and water, because their caretakers can gather only so many ants in a day—two came promptly to lap it up. Once released to the wild, they would revert by instinct to a diet of ants and termites.

THE GOAL AT Iberá to rewild Tompkins properties has included combining them with government lands (both national and provincial, some already reserved for nature) into a great public park—the big-park vision—and to nurture tourism-based economic development in communities around the wetlands' perimeter. That struggle has been long and fraught. Sofía Heinonen, executive director of Tompkins Conservation in Argentina, who started managing the Iberá project in 2005, told me that people first spoke of Doug Tompkins as "the gringo who wanted to steal the water." It became an opposition slogan: *Los gringos vienen por el agua*—The gringos are coming for the water. Argentines found difficulty—as had Chileans during the Huinay time—in believing that two rich Americans would buy land just to give it away. Some officials of Corrientes Province were also suspicious of the big-park vision, as were major local landowners, embracing the older economic model of cattle, forestry, and rice.

Support from Corrientes officials was critical because, apart from Tompkins properties and land held by the national government, much of Iberá belonged to the province. "We knocked the door, we knocked the door," Heinonen told me. Corrientes officials wouldn't open it. But mayors of the small towns surrounding the wetlands, gateways to the ecosystem, were showing more interest in the potential tourism revenue from a big park. And the national government in Buenos Aires, especially the Ministry of Tourism, also saw Iberá as a promising new destination. By 2013, at least one politician in Corrientes, Senator Sergio Flinta, realized that the province was on the wrong side of this fight and began pushing park-creation bills in the provincial senate. But it was still a deadlock. Then an event broke the impasse: Doug Tompkins died.

Immediately, amid her grief, Kris Tompkins acted. She told Heinonen to call Flinta and close the deal on compromise terms, involving 415,000 acres of Tompkins land, plus Corrientes provincial land, plus Argentine national land, all linked (but no sovereignty subsumed) to form a single great park. Within two weeks, Tompkins, Heinonen, and Flinta were in the office of Mauricio Macri, Argentina's new president, and the deal was made. Tompkins could have worn widow's black to that presidential meeting, playing on sympathy, but she turned up in a white sweater and managed a smile, expressing the implicit message: *Enough political quibbling, life is short. Let's get it done.*

Five years on, former critics had come to see both the heritage value of rewilding and the economic benefits of tourism. "There were people who didn't like Doug because he was a Yankee," Flinta told me. "And now they say thank you."

BACK IN PATAGONIA National Park in Chile, I rode up the Chacabuco Valley one day with a bird guide to view Chilean flamingos and grebes and coots and other waterfowl from an overlook above Lago Cisnes—Swan Lake—a reed-rimmed widening of the Chacabuco River. The namesakes were there too: black-necked swans, so elegantly pied, and little coscoroba swans, white-faced, with black wing tips. At the lake's west end, *álamo* trees (known elsewhere as Lombardy poplars) shaded a table and a small sign: *ÁREA DE PICNIC PICAFLOR Y ÁGUILA.* Lolo and Birdie first camped at this spot in 1993, on their way to explore Argentina, and returned to it nearly every year until his death. Today, a family of Chileans from a nearby town, with a Santiago friend on a visit, were sharing lunch at the picnic site. I spoke with the wife, a lawyer named Andrea Gómez Jaramillo. Yes, she said, we have come here before, we enjoy the wildlife, the guanacos are fun. The museum down at park headquarters is spectacular. Once, a year ago, we even saw a puma—including Renata, my daughter here, yes, she saw it too. An experience to remember.

That evening, again in the stone guesthouse on the hill, while we ate a pasta dinner Tompkins had cooked, she mentioned that she would fly off early next morning in the Husky, with her pilot, to look at an interesting place on the Chilean slopes of Cerro San Lorenzo, just south along the high Andean border, that might merit buying.

"When does it end, Kris?" I asked.

"It doesn't," she said. "Until I kick the bucket."

* * *

It hasn't ended; it has continued with new initiatives and concrete results. "Rewilding has become the big focus," according to Carolyn McCarthy, global communications director of Tompkins Conservation. "Functioning ecosystems are what we want," she told me recently. That means not just protecting habitat, in the form of land acquisitions that become national parks, but also bolstering populations of scarce and endangered species within those protected areas and, in cases where species have been extirpated entirely, reintroducing them not just for their own sake but also so they can play their ecological roles, as predators, prey, scavengers, seed dispersers, or whatever, within the intact systems. Efforts with jaguars and giant anteaters (mentioned previously), and with giant river otters,

Darwin's rheas, and red-and-green macaws, which I saw during my visits in 2019, have put animals back into the wild and been broadened to include other species.

Red-footed tortoises, rescued from the illegal pet trade, are being released from an acclimatizing facility into El Impenetrable National Park in northern Argentina. The red-footed tortoise had been extirpated from Argentina, mainly because it was captured and trafficked for the pet trade and because areas of its habitat were destroyed for agriculture. Of the 40 rescued individuals brought to El Impenetrable, the first 10 were released in May 2022, and two females were soon reported to have laid eggs in the wild. Eight jaguars, including some born and raised in the facility at San Alonso, and one adult male rescued from the Brazilian Pantanal, have been released into Iberá. Four cubs have been born in the wild. Two juvenile Andean condors were released into Patagonia National Park after being rescued from other parts of Chile and cared for in a raptor rehabilitation center. One of the birds is named Pumalín. Four female red-and-green macaws, and one male, have flown free in Iberá, and the male and one female have successfully hatched a clutch of three chicks, of whom two (at latest report) have survived. It's incremental work, nurturing such precious individuals, like blowing gently on the dry tinder of a life-or-death campfire. But with care, patient attentiveness, and good luck, the individuals become a revivified population; the campfire blazes and warms.

Tompkins Conservation itself has hatched and fledged two organizational offspring: Rewilding Argentina and Rewilding Chile, which now conduct this ongoing work. Rewilding Chile, in collaboration with agencies of the national government and some private landowners, is promoting the concept of a National Huemul Corridor, to allow those endangered Andean deer to range between patches of secure and productive habitat. And the goal of revivifying ecosystems includes the human communities, with their economies and cultures that are part of such ecosystems. Rewilding Argentina works with local communities neighboring El Impenetrable to create family gardens for vegetables, which are in scarce supply. In Iberá, Rewilding Argentina created a "community cooks" program (Cocineros del Iberá), meant to preserve local culinary traditions and offer such fare (rather than tired internationalized options such as burgers and fries) to tourists who come for the park.

Land purchase followed by land donation remains an important part of what Tompkins Conservation does. Down at Cape Froward, at the tip of the Brunswick Peninsula, which is the southernmost reach of the Chilean mainland, Tompkins Conservation and Rewilding Chile have bought another 230,000 acres and are working jointly toward creation of still another new national park. The land supports huemuls and includes coastline along the Strait of Magellan, habitat for humpback and sei whales, thick forests of kelp, and other marine biodiversity for which Tompkins Conservation and Rewilding Chile hope to help establish marine protected areas.

The Tompkins model differs dramatically from what other conservationists and organizations have done—in the Russian Arctic and Kamchatka, across equatorial Africa from Gabon to Tanzania, from Angola to Uganda, and elsewhere in wild places around the planet. Tompkins Conservation began from the circumstance of two private individuals, enormously wealthy and keen to commit large amounts of that wealth to purchasing lands for conservation, with the intent of donating those protected lands back to the citizenry of the nations in which they lie. Greg Carr's efforts at Gorongosa also began from the advantage of personal wealth, along with his immersion in thinking about both biological diversity and human rights. Other efforts have arisen from the passionate commitment of individuals (such as Mike Fay, Jane Goodall, and Steve Boyes) with no great financial resources of their own but plenty of skill, passion, and energy, and from the actions of national leaders (such as Nelson Mandela, Joaquim Chissano, and Omar Bongo) capable of transforming the vision of others into policy. Still others, such as the work of Pristine Seas, combine scientific expertise and expeditionary teamwork with diplomatic skills of high aplomb. No single approach fits every situation, and no single fund of resources (money, scientific acumen, passion, patience, political leverage, popular appeal) answers every need. But there are valuable lessons to be gleaned, I hope, from the case histories gathered here. The last one that I'll offer, the following chapter, describes an entity and an approach quite unlike anything else. Only the goals are the same.

BOOTS ON THE GROUND

•┤•┤•┤•

A Private Organization Undertakes to Protect and Run African Parks

During the last years of the 20th century, under the presidency of Nelson Mandela, South Africa began an effort to revitalize the management of its national parks. As of 1997, the CEO of its parks agency was Mavuso Walter Msimang, a colleague of Mandela's from the African National Congress (ANC) party during its years of struggle against apartheid. Msimang and others realized that an entirely new management model—even a new institution—was needed to rid parks management of corruption, inefficiency, and insolvency. In 2000, he and a group of four other men, including the Dutch philanthropist and conservationist Paul Fentener van Vlissingen and the Zimbabwean Peter Fearnhead, then head of commercial development for South Africa's parks, founded a new organization, a private nonprofit, called the African Parks Management and Finance Company. Its mission was rigorous management and protection of parks—initially in South Africa but soon, as its appeal spread, on a contract basis elsewhere across Africa. The name evolved to African Parks Foundation, then African Parks Network, and now simply African Parks.

In early 2019, photographer Brent Stirton and I were asked to do a National Geographic story on this organization. Having worked together previously on the lion-human conflict piece ("Tooth and Claw," pages 203–210), Brent and I were happy to collaborate again, and happy also for the chance to see some of the magnificent and heavily challenged national parks of which African Parks had assumed management authority. We traveled together for some of our *field time, independently for other trips. Brent went to the Ennedi reserve in remote northeastern Chad, and we shared time and a Land Cruiser at Zakouma, in southern Chad. I revisited Pendjari in northern Benin, where I had been once before while researching lions. We both got to Garamba National Park in eastern Democratic Republic of the Congo. This is the scope of travel and research for a magazine article that* National Geographic, *as a rarity among magazines in the world, made possible. I've been grateful. "Join the Navy, see the world," goes the slogan. Or work for the magazine with the yellow border.*

••

What they call the central control room at Zakouma National Park in southeastern Chad is a windowless space on the second floor of the headquarters building, a sand-colored structure with a crenelated parapet that gives it the look of an old desert fortress. Outside the room's door hangs an image of a Kalashnikov rifle, circled in red, with a slash: No weapons allowed inside. That's relevant because Kalashnikovs are almost ubiquitous in Zakouma. On a good day, only the rangers carry them. On a bad day, they are also in the hands of intruders who want to kill wildlife.

Acacias shade the headquarters compound, Land Cruisers come and go, and not many steps away on a morning in March 2019, several elephants drank from a pool, appearing almost as tame as backyard ducks. They were not tame; they were wary but thirsty. Zakouma has been a war zone for elephants since the 1970s, losing 98 percent of its elephant population, with almost 4,000 of those animals killed during just the first decade of this century, mostly by Sudanese horsemen riding in from the east on almost militarized raids for ivory. These raiders are known as Janjaweed, which sounds exotic and chilling, a word that can be translated as "devils on horseback," though some also ride camels. Their origins lie among nomadic Arab groups with high equestrian skills that, once armed and supported by the Sudanese government, served as ruthless strike forces during the civil war in Darfur; later, they became freelance bandits with a lust for ivory. For a while it seemed they might kill every elephant in Chad.

But in 2010 a private organization called African Parks (AP) took over management of Zakouma, on contract with the Chadian government, and AP's efficient new regime, including more than a hundred well-trained and well-armed rangers, stanched the flow of elephant blood. Since then, only 24 elephants have been killed in Zakouma, and no ivory lost. The Janjaweed have been repelled, at least temporarily, toward softer targets elsewhere. And the elephants of Zakouma, after decades of mayhem and terror, have resumed producing young. Their population, at the time I visited, included about 120 calves, a sign of health and hope.

The nerve center of this effort is the central control room, where fresh intelligence on elephant locations and any troubling human activity—an illegal fishing camp, a gunshot, a hundred horsemen with automatic rifles galloping toward the park—is received and used to inform ranger deployments. The sources of information include reconnaissance overflights, foot patrols, GPS collars on elephants, and handheld radios placed with trusted informants in villages around the park. The daily briefing began, on the morning of March 19, 2019, as it did every morning, at 6:00 a.m. Access was restricted to a small group of key tacticians, a radio operator, and, this day, two visitors—photographer Brent Stirton and me. A long desk held a pair of computer monitors and, on the wall behind, a large map was decorated with stickpins. Tadio Hadjibaguela, an imposing Chadian man in a turban and camo fatigues, head of law enforcement for the park, presided in French. Leon Lamprecht, a South African who grew up in Kruger National Park, and who came to Zakouma as AP's park manager, oversaw the meeting and, with quiet asides while Hadjibaguela spoke, gave Brent and me a gloss on what was being settled.

The black pins on the map represent elephants, Lamprecht said. The green pins are regular ranger patrols—known as Mamba teams—six men to a team, bushwhacking through the park on five-day rotations. "Nobody knows where they're going," he said. Their movement is unpredictable because it's dictated by the elephants, whom the Mambas follow discreetly like guardian angels. And this, said Lamprecht—pointing to a red-and-white pin set aside from the map—represents a Phantom team, two men, doing long-range reconnaissance. Those are so secretive that not even the radio operator knows their locations, only Lamprecht himself and Hadjibaguela. All the data are collated each morning, again each afternoon. "We play chess twice a day," Lamprecht said. Across the chessboard are Janjaweed and every other sort of poacher who might test the boundaries of Zakouma.

High on the wall, above the maps, hung a series of plaques commemorating the losses, low in number but deeply begrudged, since African Parks took responsibility here. *Incident. 19 December 2010. Zakouma NP. 4 elephants*, reads one. Another: *Incident. 24 October 2010. Zakouma NP. 7 elephants.* Every animal counts. Amid the row of memorials is another, different but equally laconic: *Incident. 3 September 2012. Heban. 6 Gardes.* The murderous ambush of six rangers on a hilltop called Heban, 60 miles northeast of the park, is a dark memory and an abiding incentive for vigilance within the culture of Zakouma.

It began with a poaching event, also memorialized by one of the plaques: *Incident. August 10, 2012. Heban. 6 elephants.* As soon as headquarters got word of the slaughter, a Mamba team of rangers was flown out to track the poachers, who were probably a small raiding party detached from a larger Janjaweed caravan. The rangers found them and, in an exchange of fire, killed two horses. The poachers fled, leaving behind much of their food and equipment, including axes for chopping tusks from the faces of dead elephants and, in one man's kit, military orders showing his membership in the Sudanese army. They dispersed into the bush or a tolerant village, and made themselves invisible, but without—as events would reveal—departing the area. Almost a month later, a fresh Mamba team arrived and put their camp atop the hill called Heban. On the first night, those new rangers got careless. In the early hours of September 3, it seems, all six men rose and set themselves piously to morning prayer. There was no sentry. The poachers reappeared suddenly and killed five of them. One ranger, as well as the camp cook, escaped downhill in the darkness. When another AP force reached Heban, two days later, the five bodies lay outside their tents. The killers had made off with all six ranger horses, plus their rifles and PK machine gun and radio. "For them it was revenge," one Zakouma official told me. The sixth ranger went missing and was presumed dead, but the cook, shot through his thigh and bleeding badly, struggled back to friendly ground and survived.

Framed photos of all six rangers hang in a gallery of honor at Zakouma headquarters, one floor below the central control room.

THREATS OF VIOLENT incursion, severe for Zakouma, are worse still for Garamba National Park, in the northeastern corner of the Democratic Republic of the Congo. Garamba is a magnificent anomaly, threatened and battered from all sides, like a church steeple standing above the battleground of the Somme.

African Parks has managed Garamba since 2005, on a partnership contract with DRC's Institut Congolais pour la Conservation de la Nature (ICCN). Of the 15 national parks for which AP had assumed management authority between its founding in 2000 and 2019, spread among nine countries and totaling 26 million acres, Garamba is arguably the most challenging. Its landscape is a mosaic of savanna, dry bush, and forest, harboring DRC's largest population of elephants, as well as Kordofan giraffes (a critically endangered subspecies), hartebeests, lions, hippos, Ugandan kobs, and other wildlife, and constituting the core of an ecosystem that includes three adjacent hunting reserves, in which some use by local people is permitted. Its history is fraught with warfare and militarized poaching. Its rhinos (northern white rhino, another critically endangered subspecies) were hunted to the brink of extinction. Its northern boundary is the international frontier with South Sudan, which fought hard for its independence from Sudan, in the early years of this century, and then suffered power struggles and civil war. Other war zones in Uganda and the Central African Republic lie not far away. Garamba's location, its dense forest areas, and its ivory have made it a crossroads, an enticement, and sometimes a battleground for rebel armies and other dangerous interlopers for more than two decades.

In early 2009, for instance, the Lord's Resistance Army (LRA)—a cultish rebel group out of northern Uganda notorious for their abduction of children to serve as child soldiers and sex slaves, and led by a fanatic named Joseph Kony—emerged from their refuge in the western Garamba ecosystem and attacked the park's headquarters village, burning many of the buildings and stealing three tons of stored ivory. The park rangers resisted, killing 20 LRA and losing 13 of their own men. Just a few years after that, roughly a thousand rebels in retreat from the South Sudan war flooded over the border. Meanwhile, it was all costing millions of dollars. After the last big LRA attack, with its human and dollar costs, the director general of the ICCN, Cosma Wilungula Balongelwa, worried deeply. "I had almost lost hope that things could hold on," he told me, during a ceremonial day at the park, for which he had flown in from Kinshasa. Back then, at the nadir, Balongelwa had asked the CEO of African Parks, a plainspoken Zimbabwean named Peter Fearnhead, whether AP might cut and run. "Peter confirmed to me: 'No, we won't abandon Garamba.'"

Naftali Honig is a former wildlife-crime investigator, an American with seven years of experience busting poachers elsewhere in Central Africa before he came here. At the time of my visit, he headed Garamba's research and development division, which includes intelligence gathering on human activities, ecological monitoring, and technology operations. He and his team have gotten some help from engineering labs at National Geographic and other organizations, developing new tools that might be quite useful. For instance, high-precision photo mapping that could be done at high speed from an airplane, not low speed from a drone. Or acoustic sensing that could distinguish a gunshot, deep in the park, from a breaking tree limb. "African Parks has let Garamba have a slightly experimental edge to it," Honig said, because it's such a large park, facing such severe external threats.

But ranger training for boots-on-the-ground operations was still crucial here, as at Zakouma. One morning I got a look at the process. First a tough

British adviser named Lee Elliott gave me an overview. Elliott, balding above his gray sideburns, joined African Parks after an army career of 24 years, having enlisted as a private, risen through the ranks, and served in Afghanistan. When he arrived in Garamba in 2016, discipline and organization of the rangers were poor. "There's good people here. It's just nurturing those good people," he told me. For instance, he mentioned Pascal Adrio Anguezi, a towering Congolese lieutenant who oversaw ranger training. Formerly a teacher, Anguezi was a straight arrow, the least corruptible man in the ranger force. "So he'll be lucky if he makes it to his 40th birthday," Elliott added drily. The man was so disciplined he wouldn't drink water in public. "It would be harder for us if we didn't have Pascal."

At the training ground, we met eight exhausted rangers who were just finishing a 48-hour ordeal of fitness-and-will challenges to screen them as officer candidates. A full course of drills the previous day, fitness workouts that night, little sleep, a run the next morning, and now they were dodging through the bush in four-man teams with their Kalashnikovs, two men always firing while the two others moved. At signal, a team went: fire and shout, shout and move, fire and shout, always forward, squeezing off short bursts, each man hollering to his partner and protecting him as he dashed to the next cover. At the end of each charge, the team focused their fire into a torso-shaped target on a tree.

We watched one of these drills, following behind. "It was okay," Elliott said, with the stinginess of a teacher who knew lives would be on the line. The real point here, he explained, was to see who still had grit, passion, and discipline even when exhausted.

OF COURSE, WILDLIFE populations and the tourists who come to see them (as well as the people who live in neighboring communities) benefit from the civil order that AP's firm-handed management brings. Tourism helps, directly by revenue and indirectly by cultivation of donors who support AP's work. Although at Garamba there was presently no tourism, the stabilized security situation made it thinkable again, and plans were afoot to create new facilities. Back in Chad, Zakouma offered three tiers of tourist experience: inexpensive camping and self-catering; a lodge complex of bungalows and an open-air dining room, from which guided game drives go out twice daily; and a relatively luxurious safari-tent operation called Camp Nomade, accommodating a small number of affluent guests in a remote northeastern corner of the park. Not far from Camp Nomade is a broad pan called Rigueik, which floods in wet season and retains a water hole in dry season that attracts hartebeests, kobs, African crowned cranes, spur-winged geese, and giraffes by the dozens.

Late one afternoon, Brent and I visited Rigueik with Dominique Rhoades, a young British woman studying behavior of Kordofan giraffes, of which only about 2,000 survive in the wild, more than half of them in Zakouma. We approached slowly by Land Cruiser across the cracked dry mud of the pan, spiraling incrementally toward the water, in which several giraffes stood shin-deep to drink. They are easily spooked, said Rhoades, not so blasé and indolent as lions. She currently had GPS transmitter units on eight giraffes, and hoped eventually for 25, as part of a continent-wide study with other researchers. Giraffes of the various subspecies are threatened all across Africa. And fitting them with transmitters is tricky; you don't put a collar on a neck that long and slender. The best current practice is to tranquilize an animal *briefly*, then give it the reversal drug while three burly field assistants pile gently on and prevent it from raising its head. Another person covers its face with a cloth, and a small hole is drilled into one of its ossicones, those hornlike protrusions on the head. A small GPS unit is inserted and painted over with antibiotic, then the animal is released.

By this time, 21 giraffes had gathered near the pool, drinking, staring into space, a couple of young males neck-wrestling playfully. But as Rhoades described the drilling procedure, they began sidling away toward the tree cover at the edge of the pan, almost as though they could hear her. With the sun low, on a mud pan in Chad, a giraffe casts a long shadow.

THE GIRAFFES AT Garamba, also classified within the Kordofan subspecies, have fared worse than at Zakouma, but the latest count found 55, up from 48 a year earlier, and those are the only giraffes in the Democratic Republic of the Congo. Early one morning I rode along with Achille Diodio, the young man charged with keeping track of them. Two rangers with their Kalashnikovs accompanied us, a routine precaution everywhere in the park, and soon after we entered good giraffe habitat—open savanna punctuated with acacia and other trees they can browse—Diodio spotted a long neck towering above the scrub on our right. He photographed the animal with a long lens and then, from his folder of ID photos, confirmed that it was GIR37F, an adult female, first sighted four years earlier. A year later it was fitted with a transmitter, but the transmitter stopped working, and Diodio was glad to see her now, alive and apparently well, periscoping us from a hundred yards off the road.

As we continued our giraffe circuit, scanning and talking, I learned enough about Diodio to grasp that he was just the sort of rising young talent of which African Parks needs more. He's Congolese, born and raised in a small town near Garamba, lucky to come from a family with enough money to send him to secondary school in a larger town (where he lived with a cousin), then to the University of Kisangani. More good luck, by way of a relative in Kinshasa, alerted him to the chance of a scholarship for graduate study in China. He won that, made his way to Harbin in August 2014, and a month later came the north China snows. Winter in September? "It was crazy," he told me with a laugh. In Harbin, he made his own luck. The first year went entirely to learning Mandarin, a requisite for the program—and so difficult that, after one day, he was mystified and despairing. But having already acquired Lingala, Swahili, French, English, and a bit of Kikongo, he managed Mandarin.

Four years later, with his master's degree from a good Chinese university and a thesis on Congolese elephants, he joined African Parks, first as a volunteer. They weren't slow in offering him a job. Several of AP's senior managers mentioned to me what they recognize as an urgent challenge: "capacity building," training and advancement of young African nationals into leadership positions within the organization.

Many young Congolese want to continue their educations beyond the village school, beyond the small town, Diodio told me. What's the constraint? Money, he said. A bit of initial support, a bit of family money, a loan, a gift, a scholarship. "You cannot study for free."

THE ARMORY AT Zakouma, on the ground floor of headquarters, is a locked room filled with Kalashnikov rifles, PK machine guns, and ammunition, some but not all of which has been confiscated from poachers. There are also a few tusks. "We are not a military organization," Leon Lamprecht said, while showing me this trove. "We are a conservation organization that trains our rangers for paramilitary."

Peter Fearnhead, the CEO of African Parks, said something similar during our long conversation. "People say we're a highly militarized organization. Uh, no." But is it important to provide rigorous law enforcement in the parks? "Absolutely, yes." And that's not just for the protection of wildlife, but also for people in communities round about, who may literally be subjected to rape, pillaging, and plundering, Fearnhead said, by the next wave of devils on horseback. "They recognize that it's the park that brings stability, safety, and security for them." Lamprecht called that "area integrity" when he drew me a pyramid diagram of the management tasks as AP sees them. Build the base of the pyramid with law enforcement, infrastructure, solid staff, area integrity; after that, you can advance upward to community development for local people, tourism, and ecological research.

Still, the emphasis on paramilitary ranger forces, well trained and well armed, performing rigorous law enforcement against offenders who are sometimes but not always belligerent gangs of outsiders, is not just a strength of African Parks but also a delicate public-relations issue. It had been especially delicate recently, when another organization involved in parks management, the World Wide Fund for Nature (WWF, and formerly the World Wildlife Fund), stood accused of countenancing human-rights abuses by some of its partners in antipoaching work. The accusations, published by BuzzFeed earlier that year, included a case at Chitwan National Park in Nepal, where Nepalese rangers funded and equipped by WWF reportedly tortured a suspected poacher to death. Although that occurred in 2006, according to BuzzFeed it was "part of a pattern that persists to this day."

How is African Parks different? "Our model makes us responsible for the rangers. They are *our* people," I was told by Markéta Antonínová, a Czech-born woman educated in Prague, who worked with African Parks for more than a decade. Antonínová was AP's special projects manager at Pendjari National Park, in northern Benin, when I met her; Pendjari was the last major refuge in West Africa for elephants and lions. Unlike the World Wildlife Fund, she told me, African Parks directly employs its rangers. If there must be strict discipline and martial training, she said, it's because African Parks accepts accountability for anything and everything those rangers do.

Pendjari was a recent addition to AP's management portfolio, new as of 2017, under a 10-year contract and a $23 million collaboration with the government of Benin (and its reputedly conservation-minded leader, President Patrice Talon), the Wyss Foundation, and the National Geographic Society. It's part of a transboundary complex that includes adjacent parks in Burkina Faso and Niger, and the Pendjari protected area (like the Garamba ecosystem) encompasses hunting zones along its southern and eastern edges, where local people are allowed to hunt. Antonínová and her partner, a Canadian named James Terjanian, came to Pendjari that year, he as park manager (after 15 years' experience running humanitarian projects elsewhere in Africa) and she to oversee several key activities, including law enforcement and research. Building the law enforcement capacity was an urgent challenge. From just 15 poorly trained guards when they arrived, the force had expanded to more than a hundred solid rangers.

Antonínová was a highly focused professional, formidable behind her horn-rimmed glasses, as apparent when her and Terjanian's toddler son, Andrew, clearly a beloved child, pestered her for attention during the hour set aside for our interview; she ignored the kid and signaled the nanny to do her job. Antonínová's prior experience with African Parks was broad: She was in Zakouma in 2012, when the rangers died at Heban (where "they were caught unarmed, unprepared," she recalled) and she was in Garamba when the Lord's Resistance Army came to burn the headquarters village just after Christmas of 2008 ("a very, very strange feeling"). As the Garamba situation grew dicey, she was ordered to evacuate with a few other staff in the park's small Cessna, because the park manager feared that the LRA would treat expats more grimly than local people. Antonínová sat in the plane, asking herself, Is this right? "How is it possible that, just because I'm white, I got evacuated and all these other people have to stay here?" Two days later, the LRA stormed the village, going first to the airstrip, where they burned the other plane.

Pendjari National Park presents different challenges, and I asked her about those. You don't have armed horsemen, galloping in to plunder ivory, I noted.

"No," she said. "Not yet."

Or armies marching in from their wars, raiding villages round about. "Not yet."

Pendjari faces less dramatic problems such as habitat loss along the edges, where local people grow cotton within the protected area, firewood is taken stick by stick, and fish are illegally harvested (sometimes by the truckload) from the Pendjari River. Occasionally, too, there is drama—such as recently, when a guide was found murdered and his two French tourists went missing, kidnapped by terrorists and taken across the park boundary into Burkina Faso. (The tourists were later rescued by French Armed Forces, in a nighttime raid that left four of the kidnappers and two French commandos dead.) Meanwhile, the entire transboundary complex is surrounded by stakeholders, pressures, politics, interests, and needs. Before 2017, Antonínová told me, "everything in Pendjari was based on mistrust and conflict." AP contracted to assume full management authority, while attempting to work collaboratively with all parties for the benefit of wildlife, landscape, and local people. "There is no other way," she said. It's the African Parks model. Either you trust us, she said, or you don't.

ONCE A YEAR, at the end of dry season, Garamba National Park celebrates Ranger Day, a festival of martial display and appreciation of the men (and a few women) who carry the Kalashnikovs and the responsibility for defending the park's wildlife and civil order. By that time, after months without rain, the heat is fierce and the ground is like brick and even the plants, not to mention the animals, seem to gasp. This year, there were premonitions the weather might break—a shower earlier in the week, a smell in the air—but the day began warm and clear. We assembled at the parade ground in late morning, and as dignitaries and visitors took seats beneath a marquee tent, as 200 local schoolchildren waited patiently for their moment, as family and friends sat on the grass and on plastic chairs around the perimeter, and as a hundred rangers held their positions at ease in midfield, Pascal Adrio Anguezi stood before us all, six foot five and commandingly crisp in his uniform and green beret, with a wireless microphone at his left cheek and a ceremonial sword in his right hand. He would be ringmaster today.

At 11:25, Pascal called the troops to attention. In marched a color guard of Congolese army soldiers—their berets orange, distinguishing them from the rangers—with the DRC flag. Then came a small band, blaring out an anthem with four trumpets, a tuba, cymbals, and two drums. They were shaky but intrepid, and wonderful in effect, like a group of River City kids from *The Music Man*. An army general reviewed the rangers, with Pascal at his side. By now it was hot enough that we were grateful for the electric fans on tall stands, sweeping back and forth across the gallery. Then the speeches began.

John Barrett, park manager of Garamba, also in uniform, spoke briefly in French, setting a tone of appreciation for the troops, present and otherwise. "Nineteen rangers have died in action here. We mourn them today." John Scanlon, an Australian attorney and former secretary-general of CITES but now AP's special envoy, a sort of global ambassador for the organization, touched on sustainable development for neighboring communities, and (with the WWF accusations fresh in everyone's mind) the need for antipoaching ardor to be tempered with scrupulous respect for human rights. Director General Balongwela of the ICCN, who had come from Kinshasa for this event, talked about the partnership between his agency and African Parks, and after half an hour of his remarks, a ranger in formation fainted from being cooked in the sun and was carried off. Following further speeches, the parade resumed: ranger units in drill step, four female rangers, five elderly ranger veterans, the schoolchildren in their blue-and-white uniforms, and again the band, tireless and brassy. An expert tracking dog, from the canine section, performed a demonstration with her handler. The final event of the day was a series of spirited tug-of-war heats, pitting rangers against regular army, rangers against rangers, eight men on a side, dragging one another across a scrape mark in the dusty infield at opposite ends of a two-inch-thick rope. Lee Elliott, the tough British adviser, officiated cheerily in the thick of it. The point was not a scrape in the dirt. The point was: Show us what you've got. By this time, the sky had darkened and it started to sprinkle.

The dignitaries departed in cars that pulled up to the grandstand before things really got wet. The tug-of-wars continued. The sprinkle turned into a downpour. The dust became mud, slippery and warm as axle grease. The rangers, tugging and sliding and falling and getting up to tug more, fought their hardest for inches of rope. Lee Elliott, soaked and dirty, grinned with pride as he lined them up for still another go. "If it ain't rainin', it ain't trainin'," said Naftali Honig. Then he and others, including myself, climbed into our Land Cruisers and headed to lunch. As we left, the rangers remained, only them, struggling gamely in difficult conditions, which is always the way it is.

....

African Parks now manages 22 national parks and protected areas (up from 15 when this story was published) in 12 countries, for a total area of more than 49 million acres (up from 26 million), or 76,500 square miles. Those parks include portions of 10 of the continent's 13 major biomes—distinct ecological communities characterized by their flora, fauna, soil, water, and climate. Three more parks are under negotiation to be included in the AP portfolio. The organization's work and its guiding ideas continue to receive strong support from international donors, including recent grants of $108 million from the Wyss Foundation and $100 million from the Rob *and Melani Walton Foundation. With the Walton grant, $75 million is committed to AP's endowment, meaning long-term investment in the financial stability of the organization—a crucial form of support, often difficult to get. AP also benefits from the Legacy Landscapes Fund, a joint public-private initiative of the German and French governments and several international institutions. Conservation is expensive and only effective when the organizations involved and measures adopted remain robust over time.*

In the various parks, it's not just a matter of protecting what's there, but also of revivifying what's barely there, and in some cases reintroducing what has been lost. African wild dogs (aka painted wolves) are a signal example. The species (Lycaon pictus) *is classified endangered, with only about 6,600 adults and 700 breeding pairs left in the wild. African Parks has recently reintroduced African wild dogs to Liwonde National Park and Majete Wildlife Reserve in Malawi, and to Liuwa Plain in Zambia (paralleling the reintroduction at Gorongosa, mentioned previously on page 290). AP has also translocated more than 200 zebras to Matusadona National Park, enhancing the resident population. The zebras were donated by the Bubye Valley Conservancy, a private wildlife reserve specializing in sport hunting. And to Akagera National Park in eastern Rwanda, AP brought 30 white rhinos from a game reserve in South Africa. That transfer required a Boeing 747 to carry the animals, their steel shipping crates, and supplemental food. These are weighty endeavors.*

Among the latest reports from Zakouma, in Chad, is the news that 900 African buffalo have been moved from the park proper into Siniaka Minia Wildlife Reserve, a satellite area also managed by African Parks and considered part of the Greater Zakouma Ecosystem. It may be the largest buffalo translocation ever, made feasible partly by the fact that these buffalo, unlike the white rhinos for Akagera, didn't have to be transported by air. They could be herded to their new home on foot.

And from Garamba, in the DRC, which in the past has been so troubled by waves of conflict and poaching, there is also some good news. "Today, Garamba is one of the most stable and safest places in the entire region," Peter Fearnhead wrote in his latest year-end report; "not one elephant was recorded poached in the past two years, and community programmes are generating almost a million dollars through social enterprises, of which 100% goes directly to local people." Something else flowing to local people near Garamba, thanks to AP and its donors, is electricity: A solar project has placed mini grids in two villages, from which 14,000 households now receive electrification for the first time.

It's the same simple (but long overlooked) principle that Greg Carr and his colleagues have put into operation at Gorongosa: If a national park brings tangible benefits to the people around its perimeter as well as to the wildlife and vegetation and other living creatures within, it will be cherished in the present and defended into the future. And its wildness can remain wild—its great ecological heartbeat can continue to pound—so long as we allow it to retain scale, connectivity, diversity, and its natural interactive processes. We humans, unavoidably, for better or worse, are part of that connectivity. We're animals on this planet, too, after all.

AFTERWORD

·|·|·|·

Still Beating

Finally, with hopes you'll forgive me, I'll speak personally again and in slightly more detail about something to which I alluded in the Foreword.

About 50 years ago, when I was a young writer who thought his calling was fiction, with one novel published and two rejected novels in a trunk, stalled out in a menial job, hard at work on a third novel that would also never see print, I began to gravitate, as a reader, toward nonfiction. I started to read books for which an English major—steeped in Faulkner and Hemingway and Chaucer and Milton and Pynchon and Didion and others— had never made time. I read Rousseau and Thoreau. I read Herodotus and Mary Kingsley. I read Peter Gay's history of the Enlightenment and Edmund Wilson's *To the Finland Station*. I read a couple books about Darwin. I did my best to read Descartes and Berkeley and Locke. I read Loren Eiseley and J. Henri Fabre and Annie Dillard. I read *African Genesis*, by Robert Ardrey, and *The Tree Where Man Was Born*, by Peter Matthiessen. Also, somehow, I got hold of a book titled *Serengeti Shall Not Die*, by the German zoologist Bernhard Grzimek.

I knew nothing about Grzimek and couldn't even pronounce his name. This was a paperback I had picked up somewhere, out of random curiosity, because it involved big, wild animals on a big, wild landscape, and I had recently moved to Montana, attracted by that big, wild landscape. I was unaware at the time that Grzimek's book was the companion piece to a documentary film of the same title, also written and directed by him, that had won an Academy Award in 1960. Grzimek and his son Michael made the film while doing fieldwork on animal migrations in the Serengeti. During the filming, Michael had died in the crash of a small plane he piloted. This was back at a time when Tanzania, within which Serengeti

National Park lies, was still a colonized country called Tanganyika. The book became an international best seller, though it never caught my blindered attention until years later.

The book's title derived, in part, from Grzimek's response to the death of his son. Michael had died, but the wild place they had shared and loved, with its animals and processes and magnificent scale, *must not* die. (The English word "shall" can denote either a prediction or a vow, but the original German title, *Serengeti darf nicht sterben,* was clear: "Serengeti Shall Not *Be Allowed* to Die.") It was a grieving father's promise and vow.

I can't find my old copy of the book. I can't imagine discarding it, so it's probably still with me and lost somewhere amid the vast, chaotic bookshelves that line this house I share with my wife, Betsy, and our family menagerie of three dogs, a cross-eyed cat, and a rescue python. But in its pages, wherever the thing is, resides a prophecy made by Grzimek. It's a noted and memorable passage. You could Google it. In coming decades and centuries, he wrote, people would no longer travel to view marvels of engineering, but they would "leave the dusty towns in order to behold the last places on earth" where great creatures live, relatively undisturbed, amid great landscapes.

> Countries which have preserved such places will be envied by other nations and visited by streams of tourists. There is a difference between wild animals living a natural life and famous buildings. Palaces can be rebuilt if they are destroyed in wartime, but once the wild animals of the Serengeti are exterminated no power on earth can bring them back.

The "wild animals living a natural life" carry extra significance when you know that Grzimek worked as a zoo director in Frankfurt as well as a field zoologist in Africa. The "famous buildings" and palaces "destroyed in wartime" likewise resonate with his experience rebuilding the Frankfurt Zoological Garden from ruins after Germany's defeat in World War II, which just 20 of the zoo's animals had survived.

I remember Grzimek's *Serengeti,* and I return to it now, because it was probably the first book on conservation I ever read. That happened around 1974, roughly a dozen years before I read MacArthur and Wilson's *The Theory of Island Biogeography,* which I mentioned in the foreword. Grzimek's book was not theoretical. It compelled by description and emotion, not by systematized data, math, and logic. That I read it when I did, soon after transplanting my life to Montana, makes it seem, in retrospect, more than coincidental to the changes and uncertainties I was grappling with at the time. The main uncertainty was: What sort of writer can I be, do I want to be, if indeed I can continue to be a writer at all? I suppose I ruminated on that distractedly each day, after I stepped away from my little writing table and went to my job as a waiter in a steak house or (when I was no longer able to abide that) as a bartender. I couldn't yet see, through the fog, what the answer would be. But the Grzimek book had started me in a new direction.

This book in your hands, unlike Grzimek's, is not a prophecy or a vow. I don't claim to match his grief or his gravitas. This book is more like a set of dispatches from mortal combat, but it's not a war memoir, like Michael Herr's or Doug Peacock's. It's more a series of adventure tales containing (I hope) takeaway lessons about the campaign to preserve wildness—what might work, and why. In that regard, you may have noticed a small number of salient points and recurrent motifs: passionate individuals, science, rigorous institutions, wealthy angels, and sensitivity to the needs of local people living amid or adjacent to protected lands and wildlife, all of which can prove inordinately valuable.

This book is also intended as a reminder and a goad, about the personal responsibilities each of us bears in the weighty matter of determining the future of biological diversity on our planet. We *cannot* let the wildness of the Serengeti and other such great landscapes vanish from this Earth simply because we feel comfortably distant from such wildness, and because its protection is a task for those heroic others. We need to think about the things we are doing—and we are *all* doing some things, with every meal we eat, every mile we travel, every product we consume, every house we build, every child we conceive—to diminish those features of wildness I mentioned at the start, those necessary elements for the continuation of the great heartbeat: the scale and connectivity of natural landscapes, the biological diversity within them, the processes by which those living creatures and their environment interact. We need to plan a secure future not just for ourselves and our offspring, if we have offspring, but also for the lions, bonobos, elephants, salmon, polar bears, legless lizards, and giant bees with which we share our world. We need to ponder our impacts, individual as well as collective, and find the will sometimes to constrain them. It's more than a matter of writing checks to the Jane Goodall Institute or the Audubon Society (but, by all means, do). It's personal. It's the choices we make. It's what we buy, and what we burn, and what we care about. It's where we place our feet.

I've tried to make this book something other than a gloomy diagnosis, or a howl of complaint about the losses of wildness this planet has suffered and is suffering. I've tried to offer you some enjoyment, some amusement, even some hope. As you have seen, it's a book about people as much as about places and ecological processes and other living creatures. It's meant as a paean to the hardy, smart women and men who do field biology, conservation, and the institutional work that supports those twin disciplines, including but obviously not limited to Mike Fay, Lee White, Jane Goodall, Dave Morgan, Crickette Sanz, Iain Douglas-Hamilton, David Daballen, Craig Packer, Matt McLennan, Jackie Rohen, Steve Boyes, Tumeletso Setlabosha, John and Terese Hart, Enric Sala, Greg Carr, Dominique Gonçalves, Pedro Muagura, Peter Fearnhead, Pascal Adrio Anguezi, Kris Tompkins, and the many others you've encountered in this book. I hope you have found them to be, like some of the other creatures mentioned herein, a diverse and appealing fauna. People are the problem, on this planet, but people are also the only solution.

It's late, but it's not *too* late. We still have great landscapes and great chances all over the world. The heart of the wild inheres in the scale, connectivity, diversity, and processes of great ecosystems, as I've said. So long as we humans recognize that reality, respect it, and take pains to preserve those elements through fervent and wise efforts such as I've described in this book, amid magnificent places including those featured here but others as well, that heart will continue beating.

ACKNOWLEDGMENTS

My first thanks go to all those—scientists, philanthropists, conservation managers, local people in and around the ecosystems I have visited—who aided my efforts to understand by offering answers to questions, hospitality, patience, and trust. Many of their names are mentioned in the text of these chapters.

Second, I offer deep thanks to my many and crucial colleagues at *National Geographic* magazine and the National Geographic Society. At the magazine, that has included Bill Allen, Oliver Payne, Kathy Moran, Chris Johns, Jamie Shreeve, Susan Goldberg, Gary Knell, John Hoeffel, and Kurt Mutchler, among others; and the photographers with whom I've had the privilege and benefit of working: Nick Nichols (always and of course), Ronan Donovan, Brent Stirton, Lynn Johnson, Charlie Hamilton James, Erika Larsen, David Guttenfelder, Joe Riis, Cory Richards, Pete Muller, Christian Ziegler, Amy Vitale, George Steinmetz, Robert Clark, Joel Sartore, Tomás Munita, Michael Melford, Louise Johns, and Randy Olson, among others. Thanks also to the translators who helped me in the field, including but not limited to Halima Athumani in Uganda. At National Geographic Books, my thanks to Susan Hitchcock, my sagacious and tireless editor, who helped me vastly strengthen and focus this book; Lisa Thomas, editorial director, who has believed in it; Heather McElwain, for a keen copyedit; and all others of the team that helped shape not just this book but also my previous book from NGB, *Yellowstone: A Journey Through America's Wild Heart*. I happen to live within the Greater Yellowstone Ecosystem, and for reasons that may be obvious, that book is closely related to this one.

My agent, Amanda Urban, and her team at ICM are the crucial professionals who enable me to make a living from labors of love. Gloria Thiede, my meticulous transcriber of interviews and compiler of clean texts and bibliographies, and Emily Krieger, my fact-checker and guardian of accuracy, have helped me enormously with this book as they have with others. And of course, Betsy Gaines Quammen, my wife and the other author with whom I share this household of howling canids (not painted wolves, but instead two borzois and a borzoid puppy), one long-suffering feline, and one unpresuming python, is the Lebowski's Rug of it all.

A NOTE ON NAMES

Because this book is full of geography, accurate naming of places has been important. That's tricky sometimes when place-names get translated, notably with regard to the Congo.

Two countries in Central Africa carry the label "Congo" and both, gaining independence after the bad old days of colonialism, chose the same name: République du Congo. One of them, formerly *Congo français,* lies north of the great Congo River, and its capital city is Brazzaville; the other, formerly *Congo belge* (thus called and not, say, *Belgisch Congo,* because French, not Dutch or Flemish, was the colonial language) is the vast country south of the river, which became Zaire during the Mobutu regime and then République démocratique du Congo, with its capital at Kinshasa. Some authorities translate "du Congo" as "of the Congo" and some simply as "of Congo." Even the two countries themselves, presenting their embassies in English, are not consistent. It seems a matter of taste, or an unconscious tic, as to whether "du" in these two cases is construed as a preposition plus an article, or just a preposition. Many good sources prefer the former, and so do I. Ergo, the country through which Mike Fay walked, for the first half of the Megatransect, was the Republic of the Congo, and the country in which bonobos live is the Democratic Republic of the Congo. Côte d'Ivoire drops the article entirely to become Ivory Coast. Go figure.

Taxonomy also is a matter of naming protocols. My more erudite friends in the realm of taxonomic biology remind me, with gentle scolding, that the Latinate name of a species (such as *Panthera leo* or, oh, *Haumania danckelmaniana*) is not the name of an animal or a plant. It is the name of a formal category within which a population of animals or plants (or fungi or any kind of living thing) is classified. The common name, such as lion, is the name of the animal. But this fastidious rule is very hard to observe when you're writing about relatively obscure organisms, including those thorny vines that belong to the category *Haumania danckelmaniana* and have no common name. I suppose I could propose one—such as "nasty thorny creeper vine that injures your feet if you're fool enough to walk through the jungle in sandals"—but it seems simpler, if you and my tax-onomist friends will forgive me, to use the Latinate categorical sometimes as though it names a creature. So, for instance, on my trek with Mike Fay, I see a towering *Pericopsis elata* tree once or twice in the forest; I encounter but fail to see a *Haumania danckelmaniana* vine on the ground, time and again, and it rips open my goddamn toes.

ORIGINAL PUBLICATION DATES

All the chapters in this book are based on articles originally published in *National Geographic* magazine. Here we list the original publication dates of each.

"The Megatransect I: The Long Follow," October 2000
"The Megatransect II: The Green Abyss," March 2001
"The Megatransect III: The End of the Line," August 2001
"A Country to Discover," September 2003
"Goualougo and Friends," April 2003
"The Megaflyover," September 2005
"The Spirit of the Wild," September 2005
"Family Ties," September 2008
"The Long Way Home," August 2009
"Let It Be," January 2009
"The Left Bank Ape," March 2013
"The Short Happy Life of a Serengeti Lion," August 2013
"Tooth and Claw," August 2013
"Elegy for a Lion," December 2018
"Desperate Primates," August 2020
"The Meaning of North," August 2014
"Reservoir of Terror," July 2015
"Saving the Okavango," November 2017
"People's Park," May 2019
"Gift of the Wild," May 2020
"Boots on the Ground," December 2019

RESOURCES

My primary mode of research for these stories was personal reporting, including interviewing, on the scene and among the people described in each piece. I carried a notebook, a daily journal, and a recorder, from which interview tracks were transcribed by my trusted transcriber, Gloria Thiede of Pray, Montana. If a statement is in quotes, that statement came verbatim from one of my notebooks or from a transcript; I do not re-create "quotes" from memory or from scribbled paraphrase. I also usually carried binoculars, iodine, small bandages, and duct tape, for reasons described in "The Long Follow."

My background research for each piece was considerable, and generally included a welter of scientific journal articles and technical reports, some of them buried deeply in my old files or long since gone, some of them rendered outdated by more recent publications and news (which I have drawn upon for the updates following each piece). It would take many pages to list all those original sources here. I will spare you that bibliographical compendium and offer instead a concise selection of readings, websites, and organizations that might be currently relevant for anyone who cares to explore a given topic, situation, or issue further.

The Megatransect. The best source for adding to your sense of this crazed and valuable enterprise is a boxed set of two books, a large one titled *The Last Place on Earth: Photographs by Michael Nichols* (that's Nick to us) and a smaller volume, *Megatransect: Mike Fay's Journals* (a selection, not the complete batch), published by the National Geographic Society in 2005, with assistance from several generous sponsors. The former volume includes more than 150 of Nick's extraordinary color photos of Central African wildlife and landscape, plus a foreword by me and an afterword by Mike. The latter volume contains an overview by Mike, facsimile pages and transcriptions from his notebooks, and a selection of black-and-white photos from the march by Nick. Together, these form a document unlike any other I know.

A Country to Discover. Gabon's national parks are managed by the Ministry of Water, Forests, the Sea, and the Environment, of which Lee White is presently minister. After two years of shutdown because of the COVID-19 pandemic, at least some of the parks began reopening for tourism, essential to the country's long-term economy and conservation planning, in the spring of 2022. It's a wondrous country, and they need visitors.

Goualougo and Friends. The Jane Goodall Institute does important work in many parts of the world, through its Roots & Shoots program (focused on the education of children and empowering them as conservationists) and other initiatives. They are at *janegoodall.org*. The Goualougo Triangle Ape Project (GTAP), founded by Crickette Sanz and Dave Morgan and based on their own work in that precious place, continues its efforts at helping conserve both apes and the ecosystems that harbor them, both at Goualougo itself and at the Mondika field site. The effort at Goualougo is now one of the longest continuous field studies of chimpanzees in the world. GTAP is a small organization with focused impact and wide resonance, and it deserves attention and help: *congo-apes.org*.

The Megaflyover. One of the ambitious outgrowths of this broad, visionary enterprise was the Sudano-Sahelian Initiative, which exists as a draft document and a concept in the mind of Mike Fay. "The basic premise of Sudano-Sahelian Initiative is that a well managed protected area with a broad mandate of landscape management over a large area around the protected area, can make the difference between degradation of the human condition and a stable and prosperous human condition," the document said. "This initiative spans ten countries from west to east: Mali, Niger, Burkina Faso, Benin, Ivory Coast, Democratic Republic of the Congo, Central African Republic, Tchad, South Sudan, and Ethiopia. These countries are known to be some of the most insecure and environmentally imperiled on the African Continent." Fay had identified a list of important and jeopardized landscapes of the Sudano-Sahelian belt, upon which he recommended focusing effort.

Mali: Gourma landscape
Niger: Termit landscape, WAP landscape (the W-Arly-Pendjari borderlands complex of Benin, Burkina Faso, and Niger)
Burkina Faso: WAP landscape
Ivory Coast: Comoé landscape
Benin: WAP landscape
Chad: Zakouma, Siniaka Minia, Aouk, Ouadi Rimé-Ouadi Achim, Ennedi, Bodele, and Tibesti landscapes
Cameroon: Faro, Maroua, Boubandjida landscapes
Central African Republic: Bamingui-Bangoran, Manovo-Gounda St. Floris, Andre Felix-Yata Ngaya, Chinko-Zemongo landscapes
Democratic Republic of the Congo: Garamba, Bili-Uele landscapes
Sudan: Boma-Bandingilo National Parks, Jonglei, Loelle, Southern National Park, and Yambio landscapes
Ethiopia: Gambella and Omo landscapes

It was an ambitious list, and though many of those landscapes already were or soon became sites of major conservation and parks-establishment projects by the national governments, in some cases in partnership with organizations such as African Parks, there remains a great need for the protection of landscape and wildlife and for the betterment of the lives of people in those places. Originally drafted by Fay not long after the Megaflyover and revised by him in April 2022, the document helps explain why he has cast his lot with African Parks and worked for them in places such as Central African Republic and South Sudan.

The Spirit of the Wild. There are more engaged organizations than I can mention, but one worth attentive interest is the Giraffe Conservation Foundation: *giraffeconservation.org.*

Family Ties. The work of Iain Douglas-Hamilton and his colleagues, including David Daballen, Fritz Vollrath, and George Wittemyer, continues: *savetheelephants.org.*

The Long Way Home. The Wild Salmon Center, in Portland, Oregon, is still working, along with Russian partners, to protect Kamchatka salmon and the rivers in which they swim: *wildsalmoncenter.org*. The Marine Stewardship Council *(msc.org/about-the-msc/our-history)* is a global NGO that works to curtail overfishing in all forms and to structure market incentives for sustainable fishing.

Let It Be. Michael Melford's photographs, along with the original text of this story, can be seen here: *nationalgeographic.com/magazine/article/russia-wilderness*. Tatiana I. Ustinova's book on the geysers of Kamchatka is, unfortunately, still unavailable in English.

The Left Bank Ape. For more on John and Terese Hart and the TL2 Project, visit *bonoboincongo.com*. For Lukuru Foundation: *lukuru.org/index.html*. And for Gottfried Hohmann and Barbara Fruth: *eva.mpg.de/evolution/staff/gottfried-hohmann*.

The Short Happy Life of a Serengeti Lion. Craig Packer's new book, *The Lion: Behavior, Ecology, and Conservation of an Iconic Species,* the summation of a lifetime's work, will be published in 2023 by Princeton University Press. Nick Nichols's photographs of Serengeti lions, including C-Boy, are included in his book *Wild,* a compendious selection of his lifetime's work, published in 2020 by Edition Lammerhuber (Baden, Austria).

Tooth and Claw. Visit and support the Lion Guardians *(lionguardians.org),* Panthera *(panthera.org),* and WildCRU *(wildcru.org/news/counting-cats-lion-landscapes).*

Elegy for a Lion. The black-and-white version of Nick's photo, as it appeared online with this piece, can be found on page 201 of Nick's book *Wild.*

Desperate Primates. The Bulindi Chimpanzee and Community Project is a small operation that could use help: *friendsofchimps.org/bulindi.*

The Meaning of North. Pristine Seas is a project of the National Geographic Society: *nationalgeographic.org/projects/pristine-seas.*

Reservoir of Terror. Two books have been especially valuable to me in understanding Ebola virus in particular and emerging viruses in general. One of them I mentioned in this story: *Filoviruses: A Compendium of 40 Years of Epidemiological, Clinical, and Laboratory Studies* (Springer-Verlag, Vienna, 2008), by Jens Kuhn. The other is *The Evolution and Emergence of RNA Viruses,* by Edward C. Holmes (Oxford University Press, 2009). Jens Kuhn and Eddie Holmes have also repeatedly been helpful to me by way of interviews and occasional consultations. Beyond those two sources, the scientific literature on emerging viruses is vast, and I cite much of it in the bibliographies of my 2012 book *Spillover: Animal Infections and the Next Human Pandemic* and my 2022 book *Breathless: The Scientific Race to Defeat a Deadly Virus.*

Saving the Okavango. The Okavango Wilderness Project *(nationalgeographic.org/projects/okavango),* like Enric Sala's Pristine Seas, is lodged within and supported by the National Geographic Society.

People's Park. Edward O. Wilson, who had helped inspire Greg Carr's efforts in conservation and was devoted to this project, wrote a fine book about it: *A Window on Eternity: A Biologist's Walk through Gorongosa National Park,* with photographs by Piotr Naskrecki (Simon & Schuster, New York, 2014).

Gift of the Wild. Further information is available from Tompkins Conservation *(tompkinsconservation.org),* Fundación Rewilding Chile *(rewildingchile.org/en),* and Fundación Rewilding Argentina *(rewildingargentina.org).*

Boots on the Ground. Learn more about and support African Parks: *africanparks.org.*

INDEX

ABOUT THE NATIONAL GEOGRAPHIC SOCIETY

The National Geographic Society is a global nonprofit that uses the power of science, exploration, education, and storytelling to illuminate and protect the wonder of our world.

Since 1888, the National Geographic Society has driven impact by identifying and investing in a global community of Explorers: leading changemakers in science, education, storytelling, conservation, and technology. National Geographic Explorers help bring our mission to life by defining some of the most critical challenges of our time, uncovering new knowledge, advancing new solutions, and inspiring transformative change throughout the world.

The Heartbeat of the Wild immerses us in twenty years of Explorer David Quammen's experiences in the wild, natural world that the National Geographic Society plays a part in protecting. In funding multiyear initiatives like Pristine Seas and the Okavango Wilderness Project, National Geographic is able to contribute to the preservation and protection of natural spaces while empowering local communities to do the same.

Pristine Seas was founded and led by National Geographic Explorer in Residence Enric Sala in 2008. To date, Pristine Seas has worked with local communities and governments to create 26 of the largest marine reserves on the planet, covering more than 6.5 million square kilometers.

Since 2015, the National Geographic Okavango Wilderness Project, led by Explorer Steve Boyes, has been surveying and collecting scientific data on the Okavango River system. Their data informs their work with local communities, NGOs, and governments to secure permanent, sustainable protection for the greater Okavango River Basin.

To learn more about the Explorers we invest in and the efforts we support, visit natgeo.com/impact.

ABOUT THE AUTHOR

David Quammen's 17 books include *Spillover, The Song of the Dodo, The Tangled Tree,* and, most recently, *Breathless: The Scientific Race to Defeat a Deadly Virus,* shortlisted for the National Book Award. His journalism has appeared in *National Geographic, The New Yorker, Outside, Harper's, The Atlantic,* and *The New York Review of Books,* among other magazines. He is a three-time recipient of the National Magazine Award, including for a cover story on evolution in *National Geographic.* He lives in Bozeman, Montana.

I AN ODE TO AMERICA'S FIRST NATIONAL PARK

NATIONAL GEOGRAPHIC

A JOURNEY THROUGH AMERICA'S WILD HEART

YELLOWSTONE

DAVID QUAMMEN